In memory of **Howard Bloch**

Contents

Acknowledgments

I AM GRATEFUL to the many people who kindly supplied me with feedback to individual chapters of the book, including Joshua Brand, Gilbert Brenner, Robert Brenner, Bernard Cohen, Richard Diecchio, Robert Ellsworth, John Evans, Robert Gallo, Thomas Gold, Steven Harris, Robert Hazen, Joan Kowalski, Kathryn Mc-Carthy, Robert Mendelsohn, James Metcalf, Eugenie Mielczarek, David Raup, Paulette Royt, Jonathan Simon, Clifford Sutton, Wade Taylor, and James Trefil. I am especially grateful to those who gave me feedback on the completed manuscript, including Xavier Ambroggio, Tom Faria, Maeke Ermarth, Rev. Ronald Gripshover, Jr., Chester Hasert, Nancy Murphy, and Gary Page. The contributions made by my sister, Mildred Ehrlich, were particularly useful. Any errors in the book are, of course, entirely my own responsibility.

NINE CRAZY IDEAS IN SCIENCE

1 Introduction

What's a "Crazy" Idea?

Have you ever wondered why so many of the ideas in modern science sound so crazy, and how to evaluate which of the current crop of crazy ideas might be true? This book will show you how you can sort out the more promising ideas without having to rely on the opinions of experts. As a physicist, I have always had an affinity for crazy ideas. Please don't misunderstand me. It's not that physicists are any crazier than anyone else. Despite the many unfortunate media portrayals of mad scientists you may have seen, some of us are reasonably sane. It's just that physics, by its very nature, is continually challenging the conventions of our commonsense world and revealing secrets about our universe that often seem fantastic to most people. Even physicists sometimes find their creations quite bizarre. One of the leading developers of ideas in modern quantum mechanics, Richard Feynman, used to tell students that they shouldn't worry too much if they don't understand quantum mechanics because it is so paradoxical that nobody *really* understands the subject. In fact, it's when you think you finally do understand quantum mechanics completely that you have probably got it wrong.

But, even the weirdest theories of science must pass one rigorous test or be discarded: their predictions must be in agreement with phenomena observed in the physical world. Well, at least that's the ideal. Sometimes developers of new theories find ways to modify the theories in order to keep them alive, even when their initial predictions don't work out. And sometimes scientists concoct theories incapable of being tested in their lifetime, or without the expenditure of billions of dollars to build the apparatus needed to test their ideas. (Despite the eagerness of scientists to promote employment opportunities for their unemployed colleagues, theorists do not intentionally seek ideas that will prove very expensive or difficult to test. It's just that most of the easy stuff has already been done.)

Scientists who develop crazy new theories have a strange relationship with their creations. On the one hand, they wish to promote them and convince their colleagues of the theory's validity—and possibly win fame, fortune, and respect in the process. Yet, in order to accomplish this goal, the developer of a new theory must do everything possible to prove the theory is *in*correct, find its flaws, and even make any weaknesses known to the community. Of course, this is the ideal situation. In reality, when it comes to their own pet theories, some scientists may act more as promoters than flaw-finders. But such actions often backfire among colleagues, who can be counted on to subject the new ideas to especially severe scrutiny.

Revolutionary new ideas in physics and other sciences are rarely accepted immediately by the scientific community. The high threshold against the acceptance of startling new ideas is not simply a matter of a resistance to change. The existing theoretical framework in most sciences was developed after passing many tests, and it should not be abandoned casually unless we are literally forced to do so because of conflicts with new observations. In contrast to the postmodernist view of science as a somewhat arbitrary collection of beliefs and methods, most scientists believe that science can progress to more correct conceptions of the physical universe. But, in order to progress to more valid

5 theories, scientists cannot afford to abandon working theories without compelling evidence of their deficiency.

Not all theories can be proven right or wrong—some are simply untestable, or "unfalsifiable." Here are three examples of untestable theories: (1) inanimate objects have feelings, but they have no way to communicate them; (2) faster-than-light particles exist, but they have no interaction with ordinary matter; and (3) the world is only about 5,000 years old, but it was created to look as if it were 4.5 billion years old. We may choose to believe or disbelieve such untestable ideas, but they are outside the realm of science because they are not falsifiable.

The great majority of strange ideas that are testable are simply wrong. For every crazy idea that leads to a great revolutionary breakthrough, there probably are thousands that lead to blind alleys. Unfortunately, it may be only in retrospect that we can determine in which of these categories a new idea belongs. Science is forever a work in progress, so that scientific truth is always provisional (subject to future testing by experiment). Although there is no sure way to tell if a new absurd idea is right—in fact, no scientific theory can be *proven* correct—there are questions we should ask ourselves that may help sort out the more promising ones. Some key questions follow.

How to Tell If a Crazy Idea Might Be True

1. Is the idea nutty? I make an arbitrary distinction between the categories of "nutty" and "crazy" ideas. According to my definition, crazy (also: fantastic, weird, bizarre, strange, absurd) ideas are inconsistent with scientists' present theories and may have a bizarre element to them; but unlike those in the nutty (also: flaky, wacko, loony, ridiculous) category, they are not inconsistent with the most fundamental principles of nature, such as the law of conservation of energy, nor are they incoherent or internally inconsistent.

2. Who proposed the idea? This one is tricky. Sometimes scientists in a particular field may gain a reputation for being mavericks who continually come up with oddball theories. This fact should not be a deterrent to carefully examining their ideas, unless the ideas often fall in the nutty category. Crazy new ideas sometimes come from outsiders who may bring a fresh perspective to a field. Entrenched leaders in some fields may have developed a reliance on a collection of generally accepted assumptions and rules of thumb, without any firm underlying theoretical basis. In such cases, it is important that outsiders do their homework and become aware of what is really known and not known. Only very rarely can outsiders who are complete amateurs do their homework well enough to make a contribution to a highly developed field of science.

Conversely, you shouldn't be overly impressed if the proposer of a crazy idea has eminent scholarly qualifications—even including a Nobel Prize. Sometimes Nobel laureates venture into fields far from their original area of expertise, and they may feel free to develop provocative ideas, which other less eminent (but perhaps more knowledgeable) scientists would not pursue. One infamous example would be the theory that intelligence is a genetic trait of races, and that the differences between blacks and whites on IQ tests reflect these genetic differences. This theory was promoted by William Shockley, who shared the Nobel Prize in Physics for developing the transistor. Needless to say, Shockley's expertise in physics gave him no special insights into the basis of human intelligence, although it may have given his theory more visibility than it deserved.

3. How attached is the proposer to the idea? When proposers of crazy new ideas are rebuffed by their peers, they sometimes develop obsessions about their idea and refuse to abandon them, even when proven to be incompatible with observation. The negative reaction of peers stimulates the proposer to do everything possible to prove colleagues wrong, even if it means being insufficiently critical about the merits of the idea itself. A key indicator here is the proposer's selectivity in paying great attention to facts that may support the idea, but paying scant attention to facts that refute it.

4. Does the proposer use statistics in an honest way? According to the Nineteenth-century British prime minister and novelist Benjamin Disraeli, "There are three kinds of lies: lies, damn lies, and statistics." Statistical claims are often made in support of theories that are completely erroneous, either because of deliberate falsification, unconscious bias, or ignorance in the proper use of statistics. One needs to be continually on the lookout for such misapplications of statistics. This is probably the surest way to spot crazy ideas that are wrong.

5. Does the proposer have an agenda? Some areas of science are far removed from politics, but others are not. In particular, in such areas as the environment and human health, the political biases of proposers may play a large role in how honestly they deal with a controversial idea. In such cases, the source of the researchers' funding may supply important clues as to their political biases. Proposers who are strongly motivated by political biases often put forth ideas that uniformly fall into one particular ideological category, such as liberal or conservative.

6. How many free parameters does the theory contain? Physicists sometimes say that with enough free parameters, they can "fit an elephant." The fewer free parameters a theory contains, and the more specific its predictions, the more confidence we can have in it if those predictions should come true. A theory with a great many free parameters may be able to adjust those parameters to obtain agreement with experiment, no matter what the outcome.

7. How well is the idea backed up by references to other work? Some proposers of new controversial ideas tend to cite heavily their own previous work and ignore related work done by others. Science is built progressively on the work of many scientists. As Isaac Newton wrote in 1675, "If I have seen further it is by standing on the shoulders of giants." It is not sufficient that a theorist demonstrate familiarity with other relevant work and cite it in any publication. We must also verify that the cited references, in fact, state what the proposer claims, and the degree to which it is claimed. We should be highly suspicious when the proposer of a new theory claims that others have demonstrated something, when the references cited in fact make no such claim or perhaps merely suggest it *could* be true.

8. Does the new idea try to explain too much or too little? Some crazy ideas purport to explain virtually everything in a given field, but in the process they invoke a number of new concepts or raise even more unanswered questions than they answer. A theory of everything that cannot actually calculate anything, or make definitive predictions that allow it to be tested, does not seem very promising.

9. How open are proposers about their data and methods? In many fields, particularly those where patents and potentially large sums of money are at stake, researchers may be secretive about their data and methods—at least until their results are published. Even when monetary motivations are absent, researchers may still be secretive initially so they can be sure to establish their priority involving some important discovery. But in other cases, when researchers remain highly secretive even after their results are published, they create the impression that they have something to hide, and would prefer that others not try to replicate their results.

10. How well does the idea agree with common sense? I just put this one in to trick you! Common sense, while it may be a good guide for coping with everyday life, is not a good guide for deciding whether strange theories might be true. Einstein came up with relativity theory only by rejecting many commonsense views of space and time that turned out to be inapplicable to the realm of very high speeds, of which we have no experience in our everyday life. On the other hand, certain precepts of common sense may serve as a very good guide—for example, we could ask if there is a much simpler explanation than the one given by the proposer. If someone were to claim that his ability to walk on a bed of hot coals without getting burned demonstrates the existence of some extraordinary psychic power, we should have grave doubts. The laws of physics offer a simple explanation for this phenomenon without invoking any psychic power. The idea of finding the simplest explanation for a set of facts is known as applying "Occam's razor," named for the fourteenth-century English philosopher William of Occam. If two theories purport to explain some phenomenon, it is reasonable and economical to choose the simpler of the two, *other things being equal.*

The preceding questions aren't the only tests one can use to separate the wheat from the chaff when we try to make sense of highly controversial or crazy ideas, but they are a good start. We will try to use these techniques in sorting through some of the crazy ideas in this book. A big part of the fun in going through a crazy idea is trying to figure out for oneself how likely it is to be true. With most of the ideas in the book, I do have a definite opinion, but I will try not to reveal it too soon so that you can make your own assessment. I won't reveal here what fraction of the crazy ideas I ended up supporting or opposing. However, at the end of each chapter I do give a completely subjective estimate of the probability that the idea is true. I have also provided a table in the epilogue to the book giving my rating for each of the nine ideas. You will also find listed here my subjective rating scheme of zero to four "cuckoos."

Why This Particular Set of Crazy Ideas?

Many ideas in science seemed crazy at one time but are now regarded as being settled, either having been laid to rest (as in the case of cold fusion) or firmly established (as in the case of plate tectonics, which grew out of an earlier "crazy" theory of continental drift). The boundary that separates the "settled" from unsettled controversies is probably a blurry one, as the existence of die-hard adherents of cold fusion demonstrates. Nevertheless, I wanted to explore here crazy ideas involving scientific controversies that are far from being completely settled.

Although I am a physicist, it seemed worthwhile to cover a wide range of ideas in various scientific areas, particularly issues relating to the environment and human health—two areas of great public concern. (Crazy ideas in the areas of health and the environment are covered in chapters 2 through 5, while those in the area of the physical sciences are covered in chapters 6 through 10.) The human health category had its own particular challenges. Nowadays we are used to so many reversals in what is considered good or bad for us that it was difficult to come up with something really outlandish that could also be true. Regarding possible topics in the social science area, there are many crazy ideas that one should approach only with great trepidation. The one I chose was the hot-button issue of guns and gun control. As it turned out, my analysis of the research on this particular issue put me on the opposite side from where I started out politically.

Virtually all the ideas in this book involve crazy ideas in the sciences, including the social sciences. There are lots of other crazy ideas in other areas that one could have looked at, but the nice thing about ideas in the sciences is that they can be supported or refuted by data. That's part of the point of this book: to help you develop methods—through the use of lots of examples—whereby you can learn how to test the validity of crazy ideas by carefully analyzing the data supposedly supporting them.

In the interest of full disclosure I confess I have a strong affinity for one of the nine crazy ideas: I have done some original research on the subject of tachyons—hypothetical particles that travel faster than the speed of light. Because of this research, I would be extraordinarily pleased if tachyons actually do exist, and my predictions as to how to confirm their existence should bear fruit. Thus, I cannot claim to be completely objective on that particular idea. Nevertheless, even though I may be a tachyon enthusiast, I have also tried to be a tachyon critic, searching for flaws and weaknesses wherever they exist. (In a number of cases, my opinion about an idea's validity changed more than once while looking into it.)

This book is intended for the general reader who has an interest in science. Most of the chapters require little math background, but the last two are probably the most challenging in that respect. Because the topics are essentially independent of one another, feel free to read them in any order. This book could conceivably also be useful in a college-level course on "Crazy Ideas in Science." The point of such a course would be not simply to explore the ideas—as interesting as some of them might be—but more importantly, to develop techniques to better sort out what credence to give controversial claims. Such a skill is crucial in helping an informed citizenry examine rationally the key science and policy questions that face us as a society. For the good of society, we should base policy choices on the best science available, whether or not it happens to agree with what we wish might be true. In this age of the information superhighway, it is no longer possible to take the position of the nineteenth-century bishop's wife, who upon learning about evolution, remarked: "Oh, my dear, let us hope that what Mr. Darwin says is not true. But, if it is true, let us hope that it will not become generally known." If you're ready for a wild ride through the wacky world of weird science, put on your thinking cap, and fasten your seat belt.

2 More Guns Means Less Crime

Issues of gun violence and gun control draw emotional responses from those on both sides of the debate. Although many may hold moderate opinions on the issues surrounding guns and gun violence, each side of the debate also contains its share of zealots. As in a religious war, each side of the divide seems quite certain of the rightness of its own cause and the base intentions of those on the other side. Many of those who favor gun control are likely to imagine those opposed to it as a collection of angry rednecks who care more for their supposed Second Amendment rights than for the death and carnage that the gun culture is inflicting on the country. On the other side of the divide, many of those who oppose gun control see their opponents either as bleeding-heart liberals or criminals who wish to keep guns out of the hands of law-abiding citizens. Because each side tends to distrust the motives of those on the other side, even modest steps at gun control bring out strong responses. For example, mandated trigger locks, which reduce accidental shootings involving children, and waiting periods to permit background checks are opposed by those on the pro-gun side. Pro-gun advocates note that such measures make defensive use of guns more difficult in an emergency and possibly represent the first step down the slippery slope leading to a ban on gun ownership. For some pro-gun advocates the only acceptable form of gun control is a steady aim.

Those on both sides of the gun debate can cite ample anecdotes and statistics to back up their position. The pro-gun side emphasizes the defensive use of guns in preventing victimization. The number of such defensive uses of guns varies widely according to different surveys, ranging anywhere from 80,000 to 3.6 million times per year.[1] Moreover, based on surveys by pro-gun organizations, simply brandishing a gun is sufficient to break off an attack 98 percent of the time.[2] According to the pro-gun side, gun ownership—and especially concealed weapons—discourages some criminals, since they can never be sure whether potential victims may be armed. This deterrence theory is supported by surveys of convicted felons who report they are much more worried about running into armed victims than police.[3] It is also supported by the different crime patterns in countries with high and low rates of gun ownership. In Great Britain and Canada, countries with low rates of gun ownership, "hot" burglaries (invading homes with occupants present) are three times more common (as a fraction of total burglaries) than in the United States, where citizens are more likely to be armed.[4]

Those who favor gun control see the ready availability of guns as a major contributing cause of violent crime. They are probably less likely to see the clean division between violent criminals and law-abiding citizens who can be trusted to use their guns only defensively. Instead, they worry about foolish gun-toting teenagers who may be motivated by some mixture of a genuine fear of attack by others and a desire to intimidate others or appear tough. Perceived slights, which in an earlier day might have led to a fistfight, can now lead to a shootout, when one or both parties are armed. Such potentially fatal escalations are not limited to teenagers. According to media reports, otherwise law-abiding citizens seem increasingly prone to react violently to perceived slights when, for example, they are behind the wheel of their car.

Likewise, with over a quarter of American families owning guns, a violent dispute in the home can also easily turn deadly. According to FBI statistics, 58 percent of victims were killed by family members or people they knew.[5] (It may, however, be somewhat misleading to lump together family members [18 percent of deaths] with acquaintances [40 percent], since the latter category could include anyone from a rival gang member to a drug dealer.) In any case, as the journalist Philip Cook has said, "If you introduce a gun into a violent encounter, it increases the chance that someone will die."[6]

Figure 2.1. © The New Yorker Collection 1994. Mick Stevens from Cartoonbank.com. All Rights Reserved.

16
Impact of Gun Ownership on Crime

We have briefly sketched out two competing theories on the relationship between guns and crime: one argues that more guns means less crime (through deterrence of criminals), while the other argues that more guns means more crime (through making aggression more deadly). Is it possible to decide empirically which of these two theories is correct? In reality there may be some element of truth to each theory, in which case the question is, which produces the bigger effect on the rate of violent crimes? John Lott Jr., in a widely quoted study published in the book *More Guns, Less Crime,* has made an important contribution to the debate by trying to answer the question empirically. Lott, a Yale University economist, is clearly not a disinterested researcher, and all his conclusions point in one direction, as indicated by the title of his book.

There were previous studies on the impact of gun ownership before Lott's, but they had been more limited either in population size or time period. One widely cited study, for example, used crime data from only five counties in the U.S.[7] In contrast, Lott's study looked at data for all 3,054 counties over a fifteen-year period from 1977 to 1992. Some of the earlier studies have also used a dubious methodology. For example, one study published in the *New England Journal of Medicine* cited a strong connection between the probability of homicide and gun ownership in the household.[8] The study compared the rates of gun ownership in households where there had been a homicide with a control group of matching households having the same demographics. This case-control method may be fine for conducting epidemiological studies, where one can more easily decide whether households are really matched, but homicide is not just another "disease" that stalks demographically matched households equally. It seems quite probable that whatever factors make a given household more likely to be the site of a homicide might also make it more likely to include gun owners. However, just because evil, nasty folks are more likely to own guns (and commit murder), it does not follow that gun owners are more likely to be murderers.

Worldwide, there are large variations in gun ownership rates as well as rates of violent crime from country to country. For example, many countries such as Israel, Finland, and Switzerland have high rates of gun ownership and low rates of violent crime, while many other countries have the opposite pattern. Overall, there appears to be no clear correlation either way between gun ownership and violent crime rates across the countries of the world. This should not be particularly surprising in view of the large legal, economic, and cultural differences across the globe.

In contrast, looking at variations in crime and gun ownership rates across the various states of the U.S. allows more control for extraneous variables, since the legal, cultural, and economic differences are smaller, while the differences in gun ownership and crime rates between the states are quite large. According to a 1988 CBS television survey of general election voters, gun ownership ranged from a low of 10 and 11 percent of adults in Connecticut and New York, respectively, to a high of 41 percent in New Mexico, with the U.S. average being 26 %.[9] Not surprisingly, voters who identified themselves as conservatives were found to be roughly twice as likely to report owning a gun as those who identified themselves as liberal. According to these same surveys, gun ownership has risen considerably between 1988 and 1996, from 26 to 39 percent. That increase in gun ownership may in part be explained by the 400 percent increase in the rate of violent crimes in the United States during the earlier period from 1960 to 1991, but it probably has more to do with public fear of crime than actual crime rate changes. Even though violent crime rates peaked in 1991 and are presently on a downward trend, public fear of crime and the media's reporting of violent crime seem to be higher than ever.

Surprisingly, despite claims that America has always been a gun culture, the modern-day fascination with firearms has not always been the case. According to historian Michael Bellesiles, "At no time prior to 1850 did more than a tenth of the people own guns."[10] Despite tales of the Wild West, gun ownership was especially low in parts of the countryside and the frontier, and it was not until the Civil War and the mass production of weapons that gun ownership soared.[10] In fact, in the decade following the Civil War, murder rates skyrocketed, and guns became the weapon of choice.[10]

What is the correlation between modern-day gun ownership and crime rates in the various states? In order to answer this question, it is necessary to take into account the other differences besides gun ownership between the states that could affect crime rates. Other factors that need to be taken into account include variations in arrest rates, personal income, unemployment rate, percentage of males, percentages of different age groups, percentages of different ethnic groups, and population density. A standard way of disentangling all the factors that can influence crime rates is by doing a "multiple linear regression" and looking at the correlation coefficients for each factor or variable. These coefficients then show how strongly each variable is correlated with crime rates.

What does John Lott find in his analysis of the relation between guns and crime? With the exception of rape, all other violent crime rates, including murder, aggravated assault, and robbery, appear to show a 3 or 4 percent *decrease* for each one percent increase in a state's rate of gun ownership. These correlations are said to be statistically significant at the one percent level ($p < 0.01$)—meaning that they would occur on the basis of chance less than one time in a hundred. Lott reports that all categories of *non*violent crime also show decreases with increased rates of gun ownership. None of these correlations proves that increased gun ownership reduces crime, because it is possible that there are other variables Lott neglected to include in his analysis, and correlation does not imply causation. Some of these additional variables might, in fact, be the cause of *both* the increased gun ownership and the reduced crime rates in certain states. For example, those states with a high rural population are likely to have lower crime rates as well as higher rates of gun ownership. In addition, suppose that, over time, politically conservative law-and-order citizens are attracted to cities and states with low crime rates. Since conservatives are twice as likely as liberals to own guns, this correlation would explain both the high gun ownership and low crime in certain states without any cause-and-effect relationship between gun ownership and crime rates.

Impact of "Nondiscretionary" Laws on Crime

Lott is aware that such explanations could cause the observed correlation. He therefore does not regard the main support for the alleged beneficial effect of guns in reducing crime as coming from gun-ownership data. Instead, he relies on data involving changes in the "nondiscretionary" laws, which grant permits to citizens to carry concealed weapons without a requisite demonstration of need, that is, without discretion. These types of data allow us to look at both variations in jurisdiction as well as time trends. Such data, therefore, would seem to be particularly useful for discovering any relationship between guns and crime, especially because different states passed their nondiscretionary laws during different years.[1] During the period 1977 to 1992—the time period that Lott studied—ten states passed nondiscretionary laws.[2] Twelve other states passed such laws before or after this period.

Lott uses the same linear regression method mentioned earlier to see if the rates of various categories of crime are higher or lower before the nondiscretionary laws were passed, being careful to include many other variables in the regression that could explain part of the variations with time. In fact, with these other variables included, Lott finds that a number of them, especially the arrest rate and some demographic variables, account for far more of the variations in the crime rate data than do the nondiscretionary laws, which only account for a meager one percent of the variations in crime rate.

Nevertheless, despite the tiny contribution of the laws, according to Lott, all categories of violent crime have rates that become lower after the passage of nondiscretionary laws. In all cases except robbery, the results are statistically significant at the one percent level, which would seem to support Lott's claim. In contrast, nonviolent crime rates *increase* after the passage of such laws by statistically significant amounts. Even this increase in nonviolent crime seems

[1] Had the states all passed the laws the same year, and had a peak or valley in the violent crime rate been found at that time, it could have been the result either of the laws or of some natural nationwide variation in crime trends.

[2] Florida (1987), Georgia (1989), Idaho (1990), Maine (1985), Mississippi (1990), Montana (1991), Oregon (1990), Pennsylvania (1989), Virginia (1988), and West Virginia (1989).

 to make sense in terms of a substitution effect, as criminals who are deterred from risky crimes involving confrontations with possibly armed citizens resort instead to property crimes.[3]

However, despite the above findings, a little thought will show that the real test of Lott's hypothesis is not found by looking at absolute levels of crime before and after nondiscretionary laws, but rather at their rates of change. For example, suppose robbery rates were on a downward trend, decreasing by 10 percent every year before the laws were passed, and they leveled off and remained constant for all time after the laws were enacted. In that case, robbery rates after the laws would be below the rates they were before the laws, but this situation could hardly be described as one in which the laws themselves reduced robbery rates.

In fact, just the opposite would be true: the effect of the laws in this hypothetical example was to eliminate a spontaneous downward trend, and therefore the laws increased the robbery rate from what it would have been otherwise. Moreover, any reasonable model of the impact of nondiscretionary laws would predict a change in the *slope* of the crime rates rather than a sudden change in the rates themselves. For example, if Lott's deterrence idea is correct, we would expect that deterrence should depend on the number of concealed carry permits issued, which would increase with time after $t = 0$ (the date the laws took effect). Therefore, we would not expect to see a sudden drop in crime rates at $t = 0$, but rather a gradual effect as more and more permits are issued, that is, a gradual or perhaps sudden change in slope at $t = 0$.

[3] The increase in nonviolent crime rates may conflict with Lott's previous finding that states with nondiscretionary laws had *higher* rates of nonviolent crime than other states.

Lott is quite aware of this problem, and he has also run his regressions looking not only at absolute crime rates, but also at their slope or rates of change before and after the passage of nondiscretionary laws. Here again he reports some statistically significant results for *changes* in the rates of all categories of violent crime coinciding with the passage of nondiscretionary laws. But the results are not nearly as statistically significant as before. In particular, he claims one percent statistical significance only for rape,[4] 10 percent significance for violent crime overall, and no significance at the 10 percent level for murder and robbery. A 10 percent probability is a rather low level of statistical significance, which most researchers would not regard as being statistically significant.

Violent-Crime Data Time Trends

We can obtain a better sense from graphs than from correlation coefficients of whether a trend actually changes at a certain time. Figure 2.2 shows Lott's results for the rate of robberies versus time, and figure 2.3 shows the overall rate of violent crime. Given the shape of the two curves on either side of $t = 0$, these graphs would seem to support strongly Lott's thesis that the passage of nondiscretionary laws reduces violent crime. The impact on the robbery rate appears particularly impressive. However, since appearances can be deceiving, we need to look a little deeper into Lott's graphs.

Two aspects of these graphs immediately stand out. First, the crime rate plotted on the vertical axes starts far above

[4] The crime of rape for which he finds a statistically significant correlation with nondiscretionary laws was the one violent crime category that, interestingly enough, he found was *not* correlated with gun ownership rates. This seems very odd if the deterrence theory is true.

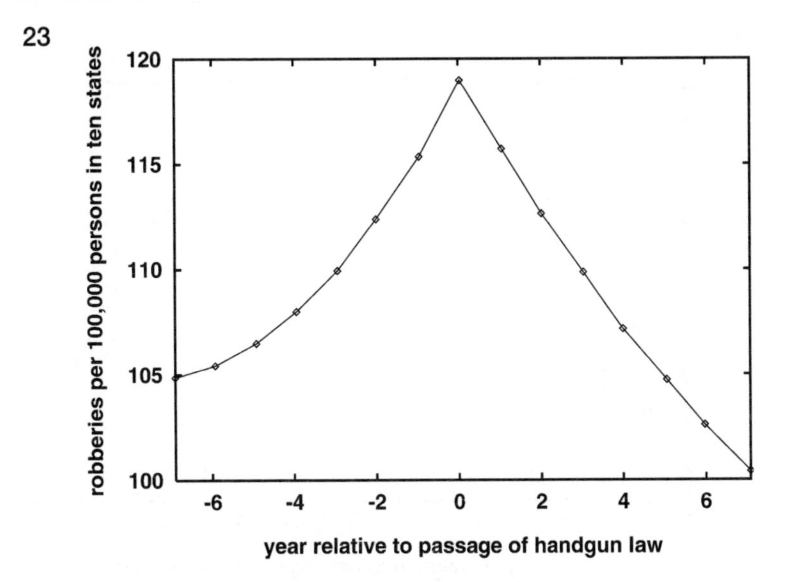

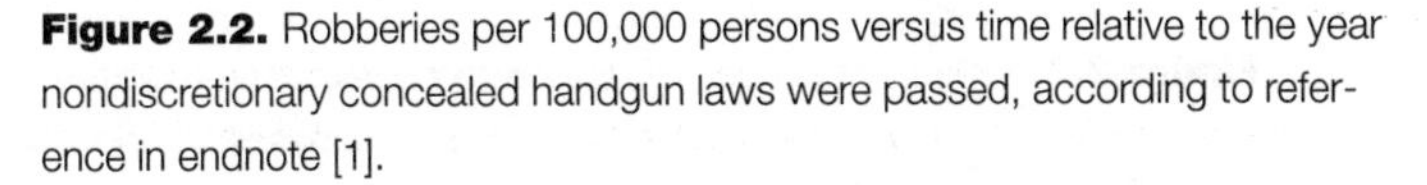
Figure 2.2. Robberies per 100,000 persons versus time relative to the year nondiscretionary concealed handgun laws were passed, according to reference in endnote [1].

zero. This is a useful technique for visually turning mole hills into mountains, or in the present case, for greatly accentuating any changes occurring at $t = 0$. Second, the points all lie exactly on the curves, because what is being displayed are not the data themselves, but merely fits to the data—a point Lott fails to mention in his book. Lott does separate parabolic fits to the data on either side of $t = 0$ and requires that the two parabolas join at this point. This highly dubious procedure almost guarantees that something interesting will happen at $t = 0$, unless the data just happen to lie on a single parabola on both sides of $t = 0$.

In order to test Lott's hypothesis, it is imperative to look at the data themselves rather than merely the fits to the data. In addition, if Lott is correct, we should be able to see the effect not merely for the data aggregated over the ten

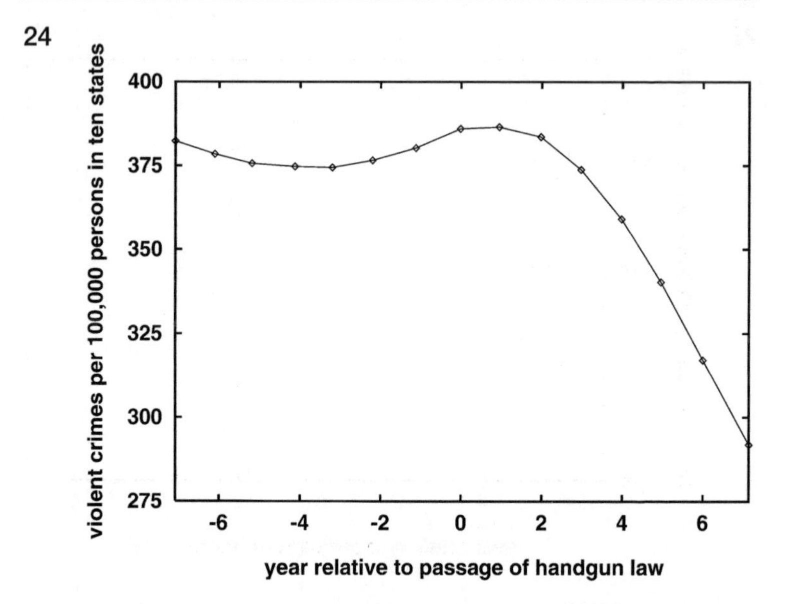

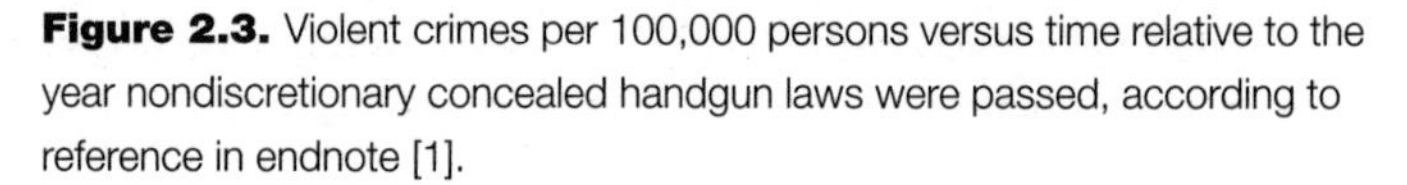

Figure 2.3. Violent crimes per 100,000 persons versus time relative to the year nondiscretionary concealed handgun laws were passed, according to reference in endnote [1].

states, but for each and every one of the states as well. In figures 2.4 to 2.6 we see plots of the robbery rates in each of the ten states, as well as for all the states combined.

We have focused on robbery specifically because it constitutes a substantial fraction of violent crimes (36.3 percent), and it was the one violent-crime category on which nondiscretionary laws seemed to show the most impressive impact, according to Lott. The data used to generate figures 2.4–2.6 were downloaded from the Bureau of Justice Statistics Web site (www.ojp.usdoj.gov/bjs/datast.htm), and were not obtained from Lott. They differ from Lott's data in several important respects, to be described later.

The main point to observe in the graphs is what changes (if any) occur for each of the ten states at $t = 0$. A close look

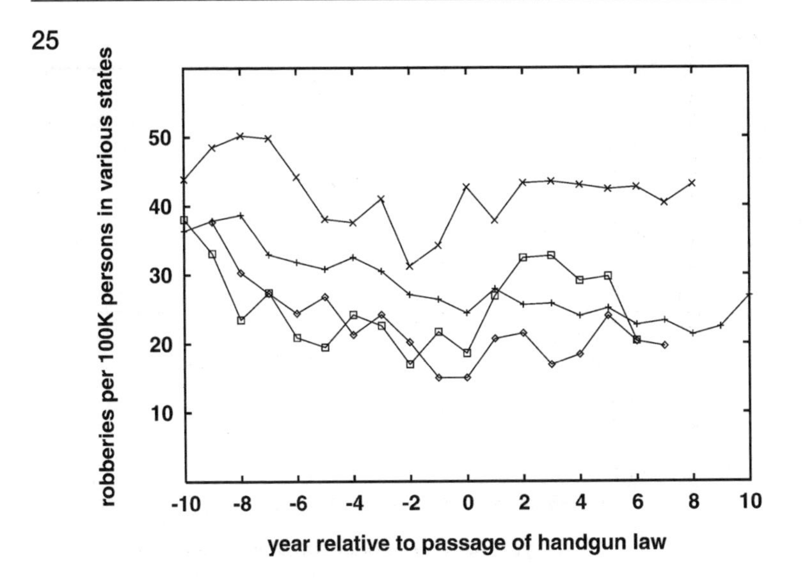

Figure 2.4. Robberies per 100,000 persons in Idaho (◊), Maine (+), Montana (❐), and West Virginia (X) in the years before and after the passage of nondiscretionary concealed handgun laws in those states.

at the graphs shows that for most of the states no significant changes appear to occur at that time. Recall that Lott showed a precipitous decrease in slope at $t = 0$ for the robbery rate (fig. 2.2), and yet of the ten states shown in figures 2.4–2.6, we find that only two of them (Georgia and West Virginia) show a decrease in slope occurring at $t = 0$, while eight of the states show *increases* in slope at $t = 0$. If we fit straight lines to the data for the four years on both sides of $t = 0$, we find that the increase in slope averaged over the ten states is $+2.4 \pm 9.9$, where the uncertainty is dominated by the contribution from the one outlier state (Georgia). An average of nine states with Georgia dropped would give an increase in slope at $t = 0$ of $+5.9 \pm 4.4$. When the data for the ten states are aggregated to show the overall robbery rate for the U.S. as a whole (see fig. 2.6), we again see a slight

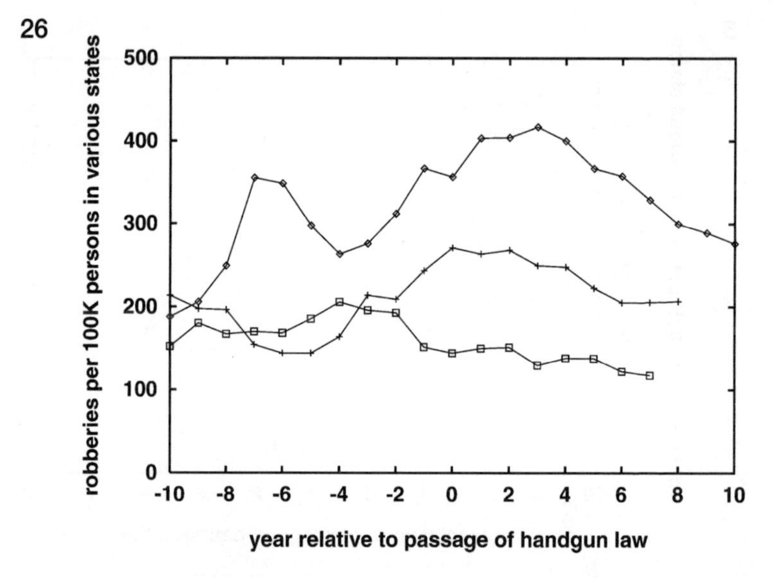

Figure 2.5. Robberies per 100,000 persons in Florida (◊), Georgia (+), and Oregon (□) in the years before and after the passage of nondiscretionary concealed handgun laws in those states.

increase in the robbery rate at $t = 0$. Thus, the impact of the passage of the nondiscretionary laws appears to lie in the opposite direction from what Lott reports, but only at a one standard deviation level of significance when one outlier state is dropped.[5]

What is the source of the disagreement between Lott's plots, which appear to show that nondiscretionary laws reduce crime, and the plots in figures 2.4–2.6, which show the contrary? First, Lott's data only went through the year 1992, while we are using data through 1997. Second, we show the actual data, rather than merely fits to the data, as Lott has

[5] If a quantity x is given in terms of a one standard deviation uncertainty range as $x \pm \Delta x$, it means that the true value of x has a 68 percent chance of falling in the interval $x - \Delta x$ to $x + \Delta x$.

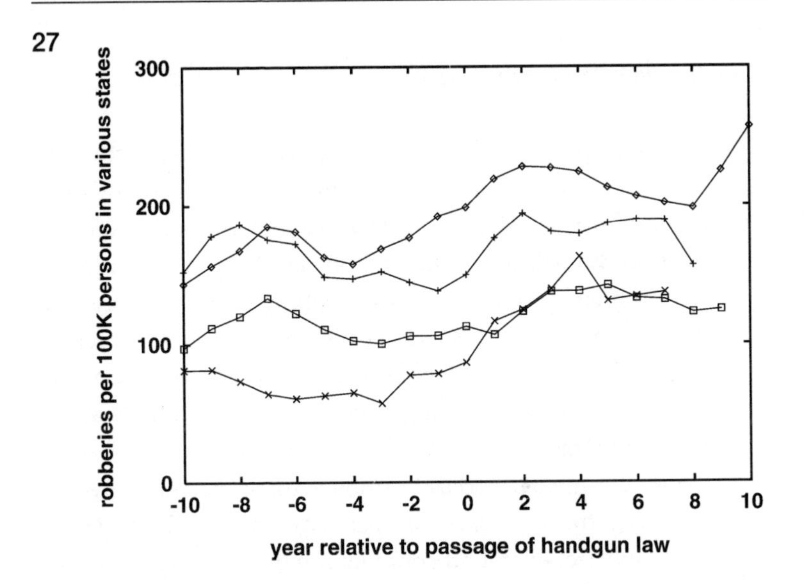

Figure 2.6. Robberies per 100,000 persons in the United States as a whole (◊), and in Pennsylvania (+), Virginia (❐), and Mississippi (X) in the years before and after the passage of nondiscretionary concealed handgun laws in those states.

done. And third, we use statewide data rather than the countywide data that Lott uses.[6] In summary, we find no evidence to support Lott's claim that nondiscretionary laws have a dramatic effect on reducing the the rate of robberies, and if anything, the opposite may be true.

[6] Lott's use of countywide data allows him to better sort out the effects of confounding variables in doing his multiple regressions. However, it is unclear why countywide data are better for spotting time trends in crime rates. If Lott were to use countywide crime rates weighted by county population, that would be tantamount to using statewide data. By not weighting the countywide data by population, however, he gives equal weight to sparsely and heavy populated county crime rates, and the average rate becomes more susceptible to the large statistical fluctuations inherent in the sparsely populated counties.

28
Time Trends in Mass Shootings

Even though mass shootings are extremely rare, in this age of media attention to violent crime and instant worldwide communication, they seem to have a disproportionate impact on the public psyche and on calls to find a solution to the problem of gun violence. Shootings involving innocent children in the public schools seem to be an especially frightening trend, to which a depressingly long series of names of towns including Littleton, Colorado, will attest. Given this high level of public concern, it would indeed be ironic if a greater access to guns actually had the effect of reducing the incidence of such atrocities, as Lott claims. (Probably not even Lott could imagine that it would be beneficial to allow kids to bring guns to school, as a colleague of mine once waggishly suggested. But we can't dismiss, out of hand, the possibility that allowing teachers to be armed might be beneficial.)

Lott defines a mass shooting as one that occurs in public places (such as schools, churches, businesses, streets, government buildings, etc.), and which results in more than one person being killed or injured. He excludes crimes involving gang activity, drug dealing, and killings that take place over a span of more than a day. Lott obtained this data through a search of news-article data bases over the period 1977–92. He argues that the same kind of deterrence argument might well explain why mass shooters might be deterred if criminals thought their potential victims might be armed. We, therefore, might expect to see lower rates of mass shootings when comparing states with and without nondiscretionary laws and when looking at the rates of such crimes, over time, before and after such laws are passed. Of course, it could also be argued that mass killers are more

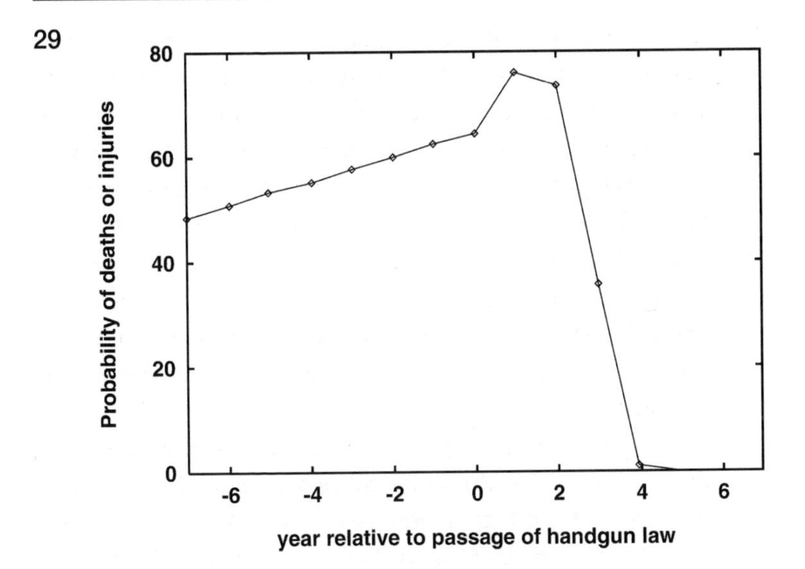

Figure 2.7. Probability that the ten states that adopted nondiscretionary concealed handgun laws during the period 1977–92 experienced deaths or injuries from a shooting spree in a public place, according to reference in endnote [1].

likely to be demented psychos who are less likely to be deterred by rational calculations of costs and benefits than ordinary robbers and muggers who are just "making a living." In any case, what does Lott find?

Comparing those states with nondiscretionary laws and those without them, Lott finds slightly more deaths and injuries in the states without the laws: 0.41 versus 0.37 deaths and injuries per million persons per year—hardly a significant difference, which probably could be explained by many factors other than the status of the discretionary laws. It is when Lott looks at trends over time that he reports a most dramatic result (see fig. 2.7).

A number of questions are raised by this impressive-looking graph, including the following: (1) Why does such a drop occur only after two years? (2) Why is the impact of deterrence on mass killers so much greater than deterrence of ordinary murderers and robbers? And (3) how can this result possibly be made consistent with his other finding that states with and without nondiscretionary laws seem to have almost the same rates of mass shootings? As with Lott's other graphs, this one "looks too good to be true" because it is. The graph, like Lott's other graphs, is based on fits to the data and not on the data themselves. In fact, figure 2.7 essentially conveys no information on the impact of nondiscretionary laws on mass shooting rates, despite the apparent precipitous drop at $t = 2$ years. One indication that the impressive-looking result shown in figure 2.7 is actually the result of random fluctuations in the data is found in a 1999 follow-up study that Lott conducted with William Landes.[11] This study covered the years 1977–95 and added the data from four additional states. Figure 2.8 is based on the data in their table 4, which shows the number of murders and injuries in mass shootings per 100,000 persons. From this graph (based on more extensive data than shown in fig. 2.7), one would have to conclude that if nondiscretionary laws had any effect on the rate of mass shootings, they somehow would impossibly have to work in reverse time—prompting a flurry of mass shootings a year before the laws were passed.

Conclusions

John Lott Jr. has made a contribution to the gun control debate by examining a more comprehensive set of data than other previous studies. His study has played an influential role in the debate, and it has provided strong ammunition to those on the pro-gun side. Despite Lott's obvious biases,

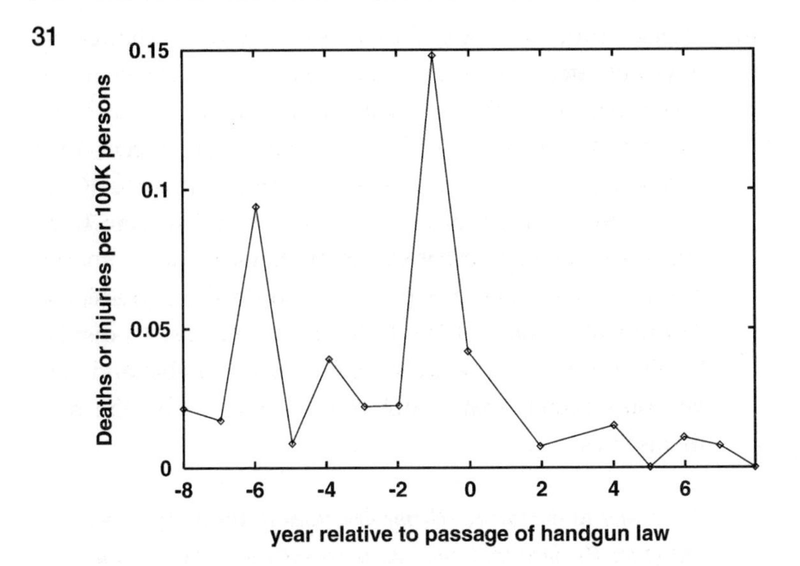

Figure 2.8. Deaths and injuries in multiple victim public shootings per 100,000 persons in the states that adopted nondiscretionary concealed handgun laws during the 1977–95 period, according to reference in endnote [11].

it is important that his study be examined carefully and not criticized for extraneous or *ad hominem* reasons. (Some critics, for example, have falsely charged that Lott's study was supported by the arms industry because he held the John M. Olin Professorship at the University of Chicago Law School.)[7] When we examine Lott's study and look at the data themselves, there seems to be little support for Lott's argument that more guns cause less crime, or that the passage of nondiscretionary laws reduces crime. The opposite also seems to be true—that is, there is no strong evidence that the increased availability of guns increases

[7] The Olin Corporation owns Winchester, the maker of ammunition. However, the Olin Foundation, while a well-known supporter of conservative causes, is completely independent of the Olin Corporation.

 violent crime, at least in the case of robberies. Although we have not attempted to duplicate Lott's regressions using his countywide data, the statewide FBI data on crime rates that cover years through 1997 (five years more recent than Lott's) appear to show no significant trend with time. (Only if we had observed an effect at $t = 0$ would it seem to be necessary to seek alternative explanations due to confounding variables, making it necessary to do the regressions.) Additionally, some of the techniques Lott uses to display his time trend data seem somewhat questionable and convey information that is not truly supported by the data themselves.

According to the rating scheme described in the introduction, my rating for the idea that more guns means less crime is 3 cuckoos. Gun control advocates will, however, be disappointed to learn that my rating for the idea that greater availability of guns increases crime is only slightly lower (2 cuckoos).

3 AIDS Is Not Caused by HIV

Most contrarian scientists can expect a chilly reception from their peers, but few are likely to provoke the level of hostility that has been directed at Peter Duesberg, who claims that the HIV virus is not the cause of AIDS. Duesberg, a professor of molecular and cell biology at the University of California at Berkeley, claims that HIV is a harmless "passenger virus," which had been around long before the AIDS epidemic began, and that AIDS is, in fact, not a communicable disease.[1, 2, 3, 4] He doesn't doubt that HIV is spread through the exchange of bodily fluids, or that the virus is more likely to be found in people with AIDS, but he claims that the HIV virus is a "surrogate marker" for the real cause of AIDS, which is the use of recreational drugs, as well as the very drugs misguidedly used to treat the disease.[1]

In fact, Duesberg essentially believes there is no such thing as AIDS as a single definable illness. In the standard view, AIDS (acquired immune deficiency syndrome) is a fatal illness consisting of a badly damaged immune system, which renders a victim susceptible to "opportunistic" infections that a normal immune system could defeat. For someone with AIDS, however, opportunistic infections can lead to fatal diseases. In fact, the diagnosis of AIDS is made based on the presence of one of around thirty specific diseases rarely seen in people with uncompromised immune

[1] A surrogate marker is a variable that is highly correlated with another variable that is the true cause. For example, poverty is linked to low SAT scores, but it is probably a surrogate marker, rather than the real cause,

 systems, such as candidiasis. The definition of AIDS has been changed several times as the specific list of opportunistic infections has expanded.[5] But Duesberg claims that the evidence is lacking that the thirty different diseases that are used to define the presence of AIDS in fact represent a single illness—that is, a compromised immune system. Rather, he believes that each of the AIDS-defining diseases has its own specific cause and is not a sign of a damaged immune system. Duesberg, a member of the prestigious U.S. National Academy of Sciences, has written at least four books[1, 2, 3, 4], and over twenty-six articles on his controversial theory.[6]

It is easy to understand why Duesberg's views have provoked such a high level of hostility among most biomedical researchers. If his claim were right, all the research, all the attempts to halt the spread of the HIV virus, and all the efforts to treat the disease represent a wild-goose chase of horrendous proportions. But the contempt for Duesberg is not simply a matter of the medical profession's distress at being told that the emperor has no clothes. To the extent that Duesberg's views are given credence, there is a real concern that large numbers of people could die needlessly if they disregard practices to prevent the spread of HIV, or forgo conventional treatments if they should become infected with the virus. Of course, if Duesberg is right that HIV is harmless, many lives are currently being lost, because efforts to stem the AIDS epidemic through stopping the spread of HIV allow the disease to spread unchecked if its real cause lies elsewhere.

A search for the cause of a new disease is similar to the process of elimination that police use to identify the person

---

which is probably some combination of poor schools, low self-esteem, and poor motivation that tends to be associated with poverty.

 who committed a crime. Sometimes in the course of a criminal investigation, the police, perhaps because of prejudice or political pressures, focus too quickly on one suspect and neglect to follow leads that might implicate others. Duesberg, in effect, claims that this scenario occurred when HIV was identified as the culprit that causes AIDS. According to Duesberg, the medical community is compounding its original error by refusing to support research that could prove the HIV/AIDS theory wrong.[7] Duesberg is joined in his controversial beliefs about the true cause of AIDS by a number of other scientists, including two Nobel laureate chemists, Kary Mullis and Walter Gilbert. In this chapter, however, we focus on the views of Duesberg, since he has been the most prolific of the AIDS-skeptics in terms of the volume of his writings and in the attention he has garnered for his views. In addition, while some of the AIDS-skeptics argue that HIV is one of a number of factors responsible for the disease, Duesberg is at the extreme end of the spectrum of opinion in arguing that HIV is a harmless virus that plays no role in AIDS.

What Is AIDS?

In order to see how much validity there is in the position of these "AIDS-skeptics," we need to consider each of the problems they have with the conventional view that HIV causes AIDS. We also need to consider how well the alternative theory (that drugs cause AIDS) fits the facts. In a foreword to one of Duesberg's books, Nobelist Kary Mullis notes that he (Mullis) began his inquiry into the cause of AIDS by seeking to find the definitive paper in the scientific literature that demonstrated that HIV causes AIDS.[1] Apparently, no one was able to cite such a paper for him, which eventually led him to conclude that a cause-effect link between HIV and AIDS has not been demonstrated. But such a conclusion oversimplifies how some scientific discoveries are made. Although many discoveries are made by a particular person at a particular time, in other cases the discovery process is more gradual and has multiple authors. A tentative conclusion may be reached initially, then becomes stronger with each subsequent investigation. This seems to have been the pattern in the case of the discovery that the HIV virus causes AIDS.

The AIDS epidemic became known in 1981 when a number of doctors in New York and California noticed a clustering of rare diseases in previously healthy homosexual men.[8] These diseases included the skin cancer Kaposi's sarcoma (KS) and opportunistic infections such as pneumocystis carinii pneumonia (PCP). One indication of the suddenness of the rise of the epidemic is that the incidence of KS rose 2,000 times (a 200,000 percent increase) in never-married men living in San Francisco from the base period 1973–79 to 1984.[9] Initially, it was suspected that the cause of the epidemic was associated with some aspect of the gay lifestyle or sexual habits, since homosexuals were the first victims.[10] However, when members of a number of other groups were found to have the syndrome, including intravenous (IV) drug users, hemophiliacs, blood transfusion recipients, and infants born to female IV drug users, those early suspicions were dropped in favor of a disease transmitted through blood or sexual contact. This suspicion was reinforced by the 1983 finding that some hemophiliacs apparently transmitted the disease to their spouses.[11]

The identification of the virus that came to be known as HIV (human immunodeficiency virus) as being the cause of AIDS was made over the course of 1983–84 by a number of researchers following on work done by American and French groups led by Robert Gallo and Luc Montagnier.[12] The initial evidence consisted of findings that nearly all AIDS patients had antibodies to the virus, that is, they were HIV-positive, while less than one percent of healthy heterosexuals had such antibodies.[13] (In one 1984 study, not one of one thousand randomly selected blood donors was HIV-positive.[14]) But there was much more besides a simple statistical association that pointed to the HIV virus as being responsible for AIDS. The virus was found to impair the body's immune system in a very specific way by infecting and killing the T-helper cells (white blood cells) that are vital for an efficient immune response to an invading organism.[15] These T-helper cells were the very cells that were found to be depleted in people with any of the AIDS-defining illnesses. Moreover, as the disease progresses, the number of T-helper cells in the blood correlates inversely with the amount of virus present—the more HIV, the fewer T-helper cells.[16]

In the conventional view of the disease, there is no sharp dichotomy between AIDS or non-AIDS; rather, there is a continuous progression from HIV infection to full-blown AIDS.[17] Figure 3.1 shows the course of the illness in a typical AIDS victim.[16] Immediately following infection with the HIV virus, the amount of virus in the blood begins to multiply, reaching a maximum in around six weeks.[16] As the virus multiples, the supply of T-helper cells becomes depleted over that same period of time. After the low point at around six weeks, the body temporarily rallies as an increasing number of antibodies to the virus are produced, and the HIV virus then proceeds to drop to very low levels, while the supply of T-helper cells partially recovers. A long latent period follows—typically lasting ten years in nonpediatric cases—in which the virus stays at very low levels,

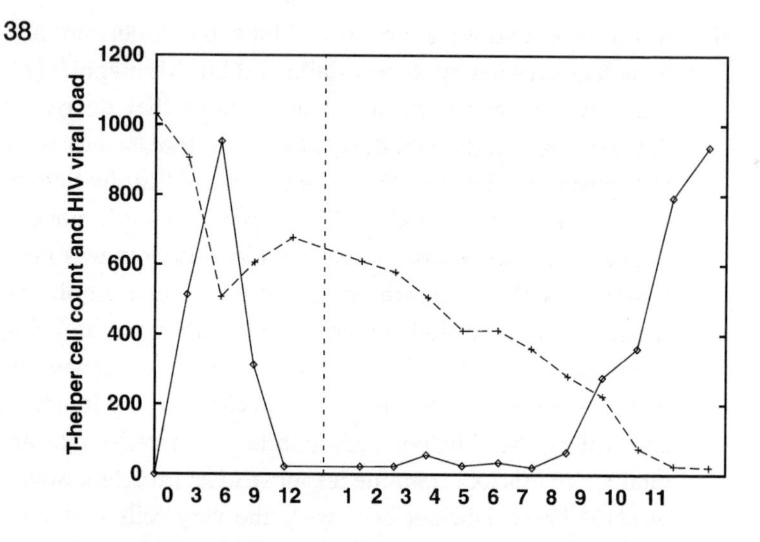

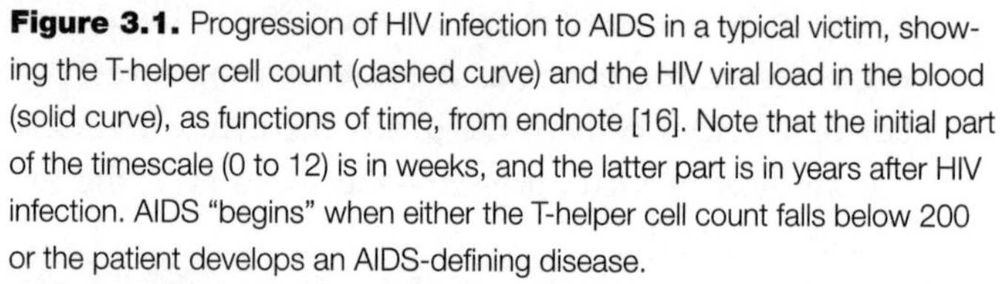
Figure 3.1. Progression of HIV infection to AIDS in a typical victim, showing the T-helper cell count (dashed curve) and the HIV viral load in the blood (solid curve), as functions of time, from endnote [16]. Note that the initial part of the timescale (0 to 12) is in weeks, and the latter part is in years after HIV infection. AIDS "begins" when either the T-helper cell count falls below 200 or the patient develops an AIDS-defining disease.

but in which the number of T-helper cells undergoes a long slow decline.[18] When the number of T-helper cells drops below around 200 per cubic millimeter of blood, the HIV virus can no longer be held in check, and the virus begins again to rapidly multiply, leading to full-blown AIDS (see fig. 3.1). In fact, in 1993 the list of AIDS-defining diseases was expanded to include a T-helper cell count below 200 in an HIV-positive individual, even in the absence of any infectious disease.[19] Given the course of the disease in a patient, doctors are able to predict when a person will develop AIDS based on monitoring the decline of T-cells and the rise of the HIV virus in the blood, all of which seem to offer compelling evidence for the conventional hypothesis that HIV causes AIDS.[20]

39 Peter Duesberg, however, finds the preceding scenario of HIV infection progressing to AIDS unconvincing, because he questions whether any virus can remain dormant in the body, only to cause disease after a long latent period in which "the active virus has been permanently eliminated from the body."[21] There is, however, no evidence that HIV has been permanently eliminated from the body during the latent period, and there are other infectious agents, such as the herpes virus that causes the skin disease shingles, which also has a long latent period. Moreover, the latent period for AIDS is somewhat of a misnomer, since latency implies dormancy. The HIV virus is simply slow acting rather than dormant during the so-called latent period, in which the number of T-helper cells experiences a long slow decline indicative of the damage HIV has inflicted on the immune system. Duesberg's objection that there really is no such single disease as AIDS, but merely thirty different diseases with different causes, is, therefore, disingenuous.

If a single virus is capable of destroying the immune system in a very specific way, i.e., by depleting T-helper cells, that is found in people with many different opportunistic infections, then (by Occam's razor—see chapter 1) it's more economical to believe that all such individuals have a common basic disease rather than thirty different diseases—unless there is some internal contradiction to that hypothesis. One contradiction cited by Duesberg is that some of the AIDS-defining diseases are not believed to be infectious, including KS, dementia, and the "wasting" disease.[22] But, as Duesberg himself points out, a virus has, in fact, recently been implicated as a possible cause of KS.[23] And just because we don't yet know of a specific organism or virus that causes a particular disease, it doesn't mean that no such agent exists, or that the disease wouldn't be more prevalent in immune-deficient individuals.

Duesberg also points to the occurrence of AIDS in certain specific risk groups as evidence that it must be a "lifestyle" disease rather than one caused by an infectious organism, which he believes would spread rapidly through the entire population, infecting males and females, heterosexuals and homosexuals in equal proportions.[23] This argument seems singularly unpersuasive; AIDS goes wherever HIV goes. The HIV virus is primarily spread through exposure to infected blood and other bodily fluids. Thus, certain behaviors, including sharing needles by IV drug users and receptive anal intercourse by male homosexuals, tend to spread the virus within those risk groups much more than to members of other risk groups. (Receptive anal intercourse is the sexual practice most likely to give someone AIDS, because in the absence of an intact condom it permits HIV in semen to enter the victim's bloodstream when small tissue ruptures occur.)

The AIDS epidemic has grown slowly in comparison to, say, a flu epidemic, only because the virus is not transmitted by casual contact, and so the number of victims grows gradually. Moreover, Duesberg's question about why AIDS wasn't evenly distributed between the sexes if it is really due to an infectious organism is being answered by the course of the epidemic, which in fact is increasingly affecting females in the United States—currently 30 percent of new infections.[25] On a worldwide basis, AIDS and HIV are found in nearly equal numbers of males and females, and mostly in heterosexuals.[26]

Does this even distribution of AIDS by gender on a worldwide basis suggest to Duesberg that the epidemic is spreading in a manner to be expected if it were an infectious disease? Not on your life. In fact, he sees the worldwide pattern as further evidence to the contrary.[23] Sub-Saharan Africa is the home to 70 percent of the world's AIDS victims, and recently U.S. officials have concluded that the impact of AIDS in some countries will be so great that it has become a threat to U.S. national security.[24] With regard to African AIDS, Duesberg notes that not only is the pattern of who has the disease different than in the West, but the range of specific AIDS-defining diseases is also different. He explains that the diseases that Africans lump under the category of AIDS are the same diseases that have long plagued that continent.[27] He suggests that calling these diseases AIDS is a convenient way to ensure the flow of money from abroad, given that AIDS draws a disproportionate share of money used to fight disease.[27]

This interpretation of the African AIDS epidemic ignores several key points, however. First, although the distribution of AIDS cases in Africa is indeed different from that in the West—the former being more evenly spread by gender and sexual orientation—these differences can be explained based on the sexual practices and greater promiscuity among Africans. Second, while the specific AIDS-defining diseases may have been present in Africa before AIDS, they previously afflicted the very old or young, or those with otherwise compromised immune systems. In Africa many more people are infected with parasites, which leads to greater T-cell activation and a resulting greater extent of viral infectivity. Moreover, since the virus has been present in Africa longer than in the West, there are more strains of HIV present. Consequently, there is no reason to be puzzled by the different set of AIDS-defining diseases in Africa and the West if AIDS is what it is purported to be—an immune deficiency—which subjects the victim to whatever infectious diseases happen to be prevalent in an area. Finally, the African AIDS epidemic is entirely correlated with the distribution of the HIV virus throughout the continent. In Malawi, the country with the highest incidence of AIDS cases, the HIV infection rate is also among the highest.[28] Conversely, on the island nation of Madagascar, AIDS is almost completely absent, and no evidence of HIV infection is found even though other sexually transmitted diseases are common.[28] Some African countries with the highest rates of HIV infection and AIDS are also among the world's poorest nations, and they cannot afford the costs of conventional AIDS drugs. It is not so surprising to find that Peter Duesberg's unconventional ideas about AIDS would have special appeal to the desperate leaders of these countries. Recently, Duesberg was appointed to an advisory board on the AIDS epidemic by the president of South Africa.[29]

What Is Necessary and Sufficient to Establish a Cause?

Establishing a cause-and-effect link between an agent and a disease is difficult to do with absolute certainty. First, let's consider the case of noninfectious diseases such as lung cancer. Most people are fairly confident that smoking causes lung cancer, even though not all smokers develop lung cancer, and not all those who develop lung cancer are smokers. In other words, the causal link between smoking and lung cancer has been established to a reasonable degree, even though the condition (smoking) is neither necessary nor sufficient for the disease to occur. Smoking is not a necessary condition for lung cancer, because the disease can occur as a result of any number of other causes, including asbestos or radiation exposure. Smoking is not a sufficient condition to develop lung cancer, because some people may not have smoked enough, or may have other unknown factors working in their favor, including, perhaps, good genes.

Infectious diseases are caused by exposure to particular microorganisms. Let's consider what is necessary and sufficient in order to show a cause-and-effect link between a microbe and a disease. Exposure to a harmful microorganism is not in itself sufficient to bring about disease, because of any number of conditions. For example, the exposed person might have good resistance to the organism, or—just as with an invading army trying to establish a beachhead—the amount or virulence of the infectious organism may be insufficient to overcome the defenses. Even though exposure to a microorganism is not sufficient to cause disease, it is considered necessary. Thus, you can't develop smallpox or tuberculosis if you haven't been exposed to the organisms that cause these diseases. Putting it differently, part of the claim that the smallpox virus causes smallpox rests on the observation that everyone who has the disease shows evidence of the virus in their bodies. Once the virus was eliminated from the population in the 1970s, the disease was effectively banished.

What about the situation in regard to AIDS? Must everyone who has AIDS have been exposed to the HIV virus in order to be confident that HIV is indeed the cause of the disease? Even though the HIV virus is an infectious organism, the necessary/sufficient criteria are not exactly the same as with other infectious diseases. Given that AIDS, by definition, consists of a compromised immune system and that many other sources of immune deficiency are known, it seems unreasonable to insist that if a single instance of AIDS is found in an HIV-negative person then the HIV virus cannot be the cause of AIDS. (Even in the case of tuberculosis there are sometimes cases where the microbe is not found.)

How often is AIDS diagnosed in HIV-negative persons? According to Peter Duesberg, there are 4,621 HIV-free AIDS cases reported in the scientific literature,[30] which, if correct, sounds like a serious flaw in the theory that HIV causes AIDS. But Duesberg's 4,621 figure involves some idiosyncratic definitions of the disease that were never intended to be definitions of AIDS.[31] A more reliable figure for the incidence of HIV-free AIDS comes from the Centers for Disease Control (CDC). One CDC survey showed that only 47 (0.02 percent) of 230,179 people diagnosed with AIDS were HIV-negative, and had persistently low T-helper cell counts, in the absence of other causes of immune deficiency.[32] Although testing for HIV antibodies is simpler than testing for the virus, sensitive tests for the presence of the virus now show that virtually all patients with AIDS do have the HIV virus in their blood.[33]

Duesberg has also claimed that it should not be surprising if 100 percent of AIDS patients are HIV-positive, because the current definition of AIDS actually requires that condition.[34] This particular criticism of the definition of AIDS has some validity, although it says little about the question of whether HIV causes AIDS. In fact, HIV-free AIDS cases were renamed "idiopathic CD4 lymphocytopenia" (ICL) in 1992, which Duesberg implies is an attempt to cover up the embarrassingly inconvenient fact of having many HIV-free AIDS cases.[35] However, whether or not such cases are called ICL or HIV-free AIDS misses the point. The main issue is that 0.02 percent HIV-free cases constitutes a minute fraction of all persons with AIDS, and therefore they do not represent the serious challenge to the conventional HIV / AIDS theory that Duesberg claims.

45
Does HIV Satisfy Koch's Postulates?

Robert Koch was a nineteenth-century pioneering bacteriologist who stipulated four tests that any disease must satisfy in order for us to be sure that it is truly associated with a particular infectious microorganism. Koch's postulates require the following:

1. The microorganism must be found in all cases of the disease.
2. It must be isolated from the host and grown in pure culture.
3. It must reproduce the original disease when introduced into a susceptible host.
4. It must be found in the experimental host so infected.[36]

Koch's postulates were originally intended to apply only to diseases caused by bacteria, not to viruses that cannot reproduce outside a host cell. Even with bacterial infections, Koch's postulates may not be 100 percent satisfied in all cases. Let's see what the situation is with regard to HIV. Koch's first postulate has already been dealt with in the previous section, where we noted that evidence of HIV infection is found in virtually all (99.98 percent) AIDS cases. Although that falls short of the 100 percent suggested by Duesberg's formulation of Koch's first postulate, recall that AIDS is a deficiency in the immune system, and that unlike the case with other infectious diseases, a compromised immune system can have causes other than the HIV virus.

How about Koch's other three postulates? Here again, Duesberg's claim that the conventional HIV/AIDS theory does not satisfy Koch's postulates has been overridden by the course of events. Koch's postulates 2–4 have been demonstrated by three laboratory workers (with no other known risk factors for AIDS) who developed AIDS after being accidentally exposed to the HIV virus in cultured form.[37] Koch's postulates are further demonstrated by health-care workers in the United States who contracted AIDS after being occupationally exposed to the HIV virus (of forty-two exposed, seventeen contracted AIDS).[38]

Duesberg, not surprisingly, has offered rebuttals to all these findings. He notes that seventeen cases of AIDS after millions of contacts with AIDS patients hardly seems indicative of a communicable disease.[39] But, of course, such rarity could just as well be indicative of the difficulty of transmitting the virus and the care taken by health-care workers to avoid exposure. (Even a needle stick with infected blood rarely transmits the HIV virus.) Duesberg also questions whether the seventeen (out of forty-two HIV-exposed) health-care workers who contracted AIDS might not have actually acquired the disease through non-HIV risk factors, such as through IV drug usage.[40] Individuals might be reluctant to admit to such behavior, so there's no proof one way or the other in any *individual* case. But even in the highly unlikely event that 50 percent of the forty-two HIV-infected health-care workers happened to be IV drug users, seventeen out of twenty-one (80 percent) is an extremely high proportion of randomly chosen persons to contract AIDS within some years of being stuck by a needle—assuming their HIV exposure really was irrelevant, and the disease was contracted based on the increased risks associated with being an IV drug user.

Moreover, the possibility of contracting AIDS due to non-HIV risks is not present in the case of laboratory animals, and Koch's postulates have now also been confirmed there as well. In one study, baboons were infected with a variant of the HIV virus and were found to suffer significant declines in their immune system, similar to that seen in AIDS patients.[41] They suffered some of the same kinds of diseases (a form of pneumonia, lesions similar to KS, and the wasting syndrome) seen in humans with AIDS.[41] While animals subject to such experiments could be argued to undergo a certain amount of stress (which might affect their immune system), it strains credulity to believe that they would suffer the same kinds of diseases and the same immune system impairment seen in human AIDS. (Animals in many other experiments unrelated to HIV could also be said to undergo a great deal of stress, and of course they never get AIDS.)

What About AIDS in Hemophiliacs and Children?

The course of the AIDS epidemic in hemophiliacs is a particularly clear piece of evidence for the conventional HIV/AIDS theory since the health of these individuals is regularly monitored, and also because many were infected with the HIV virus around the same time. Tests of the U.S. blood supply show that in 1978 at least one batch of clotting factor VIII became contaminated with the HIV virus.[42] As a result, 2,300 persons received that batch of contaminated blood, and the first AIDS cases were reported in U.S. hemophiliacs four years later.[43] Such a delay is consistent with the conventional theory, since in nonpediatric cases, the illness is found to have a latent period which is typically ten years, although it can be longer or shorter.

Figure 3.2 shows the death rate among British hemophiliacs with severe hemophilia.[44] Note that the death rates for HIV-negative hemophiliacs remained constant over the interval 1978–92, but the death rates soared dramatically for HIV-positive individuals beginning around 1985.[44] The

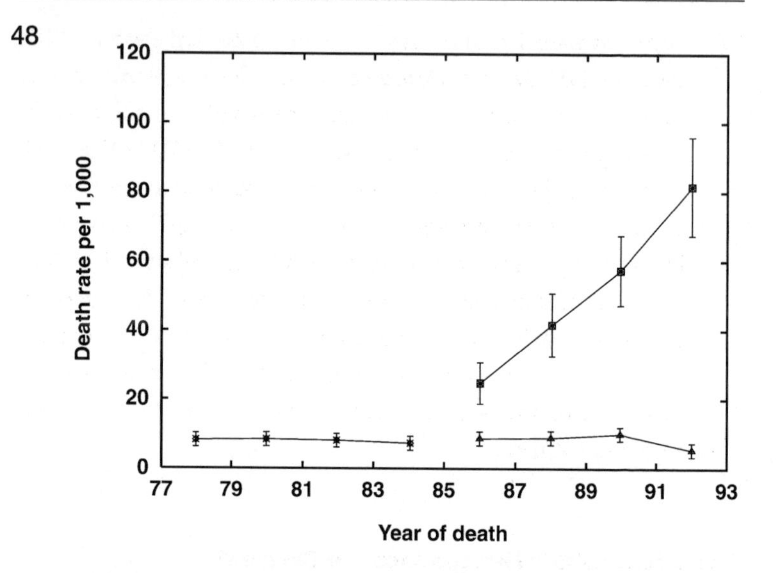

Figure 3.2. Death rate for British hemophiliacs whose hemophilia is severe. Data before 1985 is for all patients; after 1985 the HIV-positive and HIV-negative cases are shown separately. Data are from reference in endnote [44].

same is true for patients whose hemophilia was mild/moderate rather than severe.[44]

This dramatic evidence for a link between HIV and AIDS is essentially ignored by Duesberg, who egregiously minimizes the impact of the AIDS epidemic on hemophiliacs by noting in a 1998 article that "Contrary to predictions of transfusable AIDS, the lifespan of American hemophiliacs has increased more than two fold from 11 years in the early 1970's to 27 years in 1987 (the year AZT was introduced)."[23] Duesberg blames the skyrocketing death rate of hemophiliacs since that time on the toxic effects of AZT—the very drug used to treat AIDS.[23] A major problem, however, is that, as figure 3.2 shows, hemophiliacs started dying in much greater numbers in 1985—two years before AZT came into widespread use.

In support of the view that AIDS is not a contagious disease, Duesberg also claims that the wives of HIV-positive hemophiliacs have only had the normal background level of AIDS-defining diseases. But this 1996 claim is based on a 1984 report by Kreiss[45], which was early enough in the epidemic (and close enough to the first contamination of the blood supply) that most wives who did become HIV-positive would not yet have had enough time to develop AIDS. In fact, according to more recent CDC statistics, there have been 935 cases of AIDS among spouses whose only risk factor was having sex with an HIV-infected partner who acquired the disease from blood transfusions.[46] When, as in this case, the statistics do not support Duesberg's theory, he resorts to anecdotal evidence and cites specific celebrities, such as Arthur Ashe and Rock Hudson, who failed to transmit the disease to their partners.[23] But, of course, such examples are entirely irrelevant, since there are many reasons why an HIV-infected person might not transmit the virus to his partner, including luck, precautions, or high resistance on the part of the partner. The question of whether AIDS is a contagious disease hinges only on whether there are some cases in which it is transmitted from person to person, not on whether we can find some cases where it is not.

Pediatric AIDS is another category that offers strong support for the HIV-AIDS theory, since babies have no behavioral risk factors. In the United States, over 7,000 children have been diagnosed with AIDS, and about 80 percent of them were born to mothers who were IV drug users and were infected with HIV primarily by sharing needles.[46] Could it be the mother's drug use, rather than her HIV-status, that determines whether babies contract AIDS? Studies have consistently shown that not to be the case: only approximately 25 percent of babies born to HIV-infected mothers are themselves HIV-infected, and it is only this group of babies that develops AIDS.[47] The other 75 percent of infants who are HIV-free never develop AIDS.

Is it conceivable, as Duesberg claims, that HIV infection in infants might be a surrogate marker for the number of IV drugs the mother took before giving birth—that is, HIV-infected babies are more likely to have mothers who took more drugs than HIV-free babies? This possibility is not consistent with studies of twins born to HIV-infected mothers in which only one twin is HIV-infected. In such cases, only the HIV-infected twin goes on to develop AIDS.[48]

Could Drugs Be the Cause of AIDS?

Anyone seeking to refute a hypothesis, such as that HIV causes AIDS, need not propose an alternative hypothesis. Nevertheless, if we can suggest an alternative hypothesis that fits the facts much better, our critique becomes that much more potent. Conversely, if we put forward an alternative hypothesis that fits the facts more poorly, and if we use much looser standards of evidence for our favored hypothesis, we raise serious questions about our objectivity and motivation.

In support of his theory that recreational drugs have brought on the AIDS epidemic, Duesberg notes that the epidemic occurred during the same period that use of recreational drugs greatly increased. Figure 3.3, taken from a 1998 Duesberg article, does indeed show that in Europe, for example, there were simultaneous increases in AIDS cases and drug deaths during the decade of the 1980s. But the graphs of figure 3.3 actually undermine Duesberg's case rather than support it. Drug deaths in Europe rose by roughly a factor of two during the last half of the 1980s, after

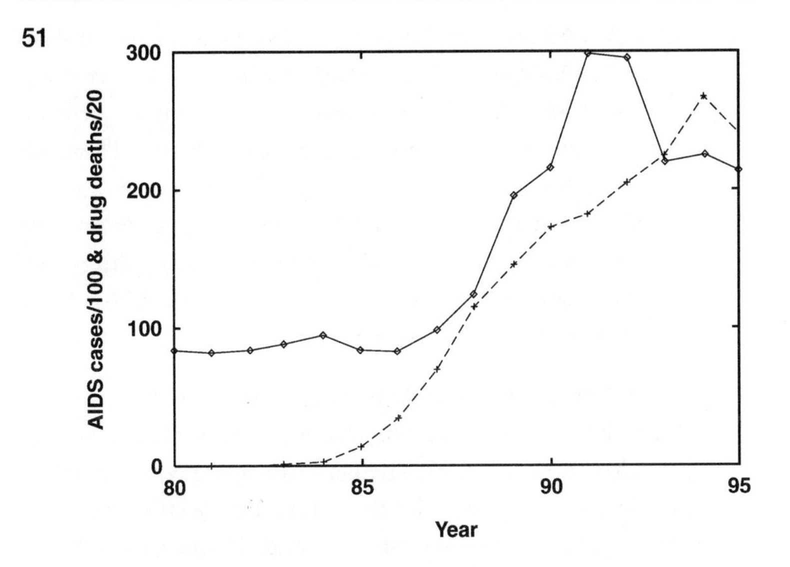

Figure 3.3. AIDS cases (dashed curve) and drug deaths (solid curve) in Europe during the period 1980–95, according to reference in endnote [23].

being fairly constant during the first part of the decade. In contrast, AIDS cases rose from essentially zero before 1985.

The complete absence of AIDS cases in the first half of the decade does not fit the theory that drugs cause AIDS. That absence does, however, fit the standard theory that HIV causes AIDS. For example, based on testing stored blood in connection with hepatitis-B studies, it has been estimated that the virus first entered the U.S. population in the late 1970s.[49] *A similar chronological association between the first appearance of HIV and the later appearance of AIDS has been noted everywhere AIDS cases have been reported.*[50]

Peter Duesberg has formulated his claim that drugs cause AIDS in very specific terms:

> The long-term consumption of recreational drugs, such as cocaine, heroin, nitrite inhalants, and amphetamines, and prescriptions of anti-HIV drugs, such as AZT, cause all AIDS dis-

> eases in America and Europe that exceed their long-established, national backgrounds, i.e., > 95%. Chemically distinct drugs cause distinct AIDS-defining diseases; for example, nitrite inhalants cause Kaposi's sarcoma, cocaine causes weight loss, and AZT causes immunodeficiency, lymphoma, muscle atrophy, and dementia. Hemophilia-AIDS, transfusion-AIDS, and the extremely rare AIDS cases that appear in the non drug-using population reflect the normal incidence plus the AZT-induced incidence of these diseases under a new name.[1]

What is the basis of Duesberg's assertion that specific AIDS-defining diseases are caused by specific drugs, and how does he know which drugs to associate with a particular disease? Let's consider the claimed association between Kaposi's sarcoma and nitrite inhalants. Nitrite inhalants are used almost exclusively by homosexual men to facilitate anal sex, and KS is also found almost exclusively in male homosexual AIDS cases, hence the presumed linkage. Early in the AIDS epidemic, before the discovery of the HIV virus, nitrite inhalants (and other drugs) were considered a possible cause of the disease, but this possibility was examined and refuted in a number of case-control epidemiological studies.

In one 1993 study, Michael Ascher et al. looked at 1,034 San Francisco men and considered their relative risk of contracting AIDS and specifically KS, based on their HIV status and the extent of their nitrite usage.[51] Ascher did find that nitrite users were at increased risk of KS and AIDS generally. The heavy nitrite users had an 83 percent higher risk of contracting KS than those who used it a little or not at all. However, such an increased risk would be expected if nitrite use is associated with receptive anal sex, which is a primary way of becoming infected with HIV. Thus, in the standard interpretation of AIDS, nitrite use is a marker or

53 surrogate for the real cause of the disease—just the reverse of Duesberg's claim.

Fortunately, there is an obvious way to tell which variable (HIV status or nitrite usage) is the real cause and which is the surrogate marker. No KS (or other AIDS diseases) were found in Ascher's study in HIV-negative men, even if they were heavy nitrite users, but sixty-six KS cases were found among HIV-positive men, even if they were light nitrite users. Duesberg, who is well aware of this study, has strongly implied that Ascher has made up some of his data and used spurious methods to analyze them. However, similar negative results for a nitrite-KS link and more generally a recreational drug-AIDS link have been found by other research groups.[52]

To be extremely conservative, however, suppose we were to assume that the excess 83 percent risk for KS among nitrite users found in Ascher's study is real, even though, as we have seen, there is every reason to doubt it. Let's further make the extremely conservative assumption that before the AIDS epidemic, no male homosexuals used nitrites, and after it began they all did—even though nitrites were widely used in the 1960s long before the AIDS epidemic. Duesberg's theory that nitrite usage causes KS would then predict that, with the onset of the AIDS epidemic, KS would rise 83 percent above its normal background pre-AIDS level. In fact, as was noted earlier, it rose 200,000 percent among unmarried men living in San Francisco.

One further issue concerning KS needs to be addressed: Why should it occur much more frequently in homosexual men than in other AIDS victims? In the context of the conventional theory, one plausible explanation is that whatever virus is responsible for the disease is transmitted most easily by the same risky sexual practices that put homosexuals at increased risk of becoming HIV infected.

AIDS is, of course, an incurable disease, but according to the conventional theory, AZT and other drugs have been useful in prolonging the lives of those infected with HIV by keeping the virus in check. Peter Duesberg claims that, on the contrary, AZT and other drugs used to fight HIV / AIDS are actually causing AIDS in otherwise healthy HIV-positive individuals. In fact, Duesberg cites one study in which AZT increased the risk of AIDS in HIV-positive individuals by 440 percent relative to those not taking the drug.[53] However, that research was not a controlled study to test the effectiveness of AZT, and no such claim was made by the authors. The authors of the study specifically noted that the higher AIDS risk of those taking AZT was in all likelihood due to the drug being administered first to those whom doctors considered to be at highest risk of AIDS, presumably based on their T-cell count.

When AZT first underwent clinical trials before widespread use, its effectiveness was demonstrated in slowing disease progression and prolonging life in double-blind controlled studies.[54] Another controlled study of AZT, however, showed no significant effect in preventing disease progression or increasing survival.[55] On the other hand, no controlled study has shown AZT to be harmful to patients.[56] One indication that AZT has not been particularly helpful (or harmful) in the fight against AIDS can be gleaned from a look at the numbers of AIDS cases and deaths versus time (see fig. 3.4). Notice the absence of any change in slope in either curve occurring in the years AZT first came into widespread use (1986–87).

One area in which AZT has proven quite effective, however, is in reducing the chances of HIV infection in babies

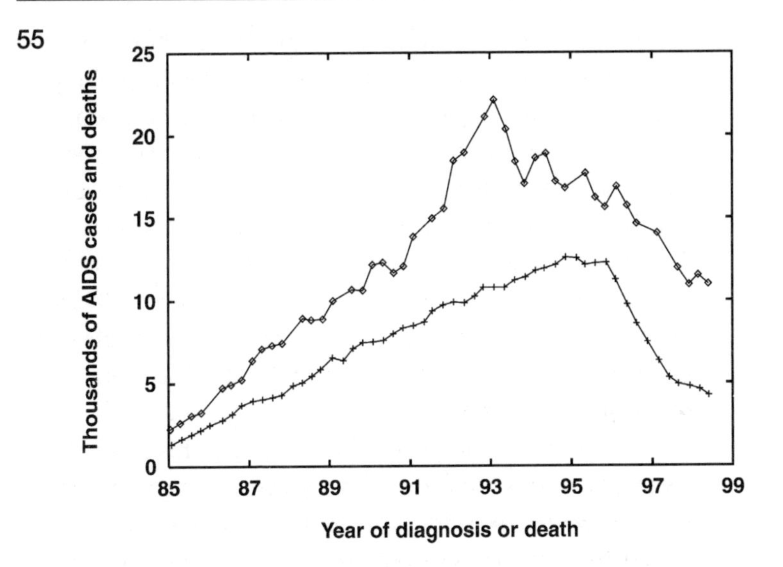

Figure 3.4. AIDS cases (upper curve) and deaths (lower curve) in the United States, according to the Centers for Disease Control during the period 1980–99.

when it is administered to the mother prior to giving birth. The chances of a mother who takes AZT giving birth to an HIV-infected baby are reduced from 25 percent to 8 percent. Here is how Peter Duesberg reacts to this more-than-three-fold reduction in risk: "Although the risk to such children of picking up HIV from their mothers is only 25%, . . . it is only reduced by AZT from 25% to 8% ."[23] One cannot help but wonder if Dr. Duesberg would refrain from taking a medication that "only" reduced his risk of contracting a deadly disease threefold, from 25 percent to 8 percent.

In recent years, a "cocktail" of three drugs has been used in the fight against AIDS, and people with AIDS are now living longer than before—as indicated by the growing gap between the number of AIDS cases and deaths shown in figure 3.4. However, the degree to which this good news can be attributed to the new drugs is open to question, since the death rate seems to have started its descent already the year before the new drugs came out in 1996. The decline in the number of new AIDS cases that began in 1992 can probably be explained most simply in terms of the decline in the number of new HIV infections that began some years earlier, that is, in 1985 (based on records of the number of HIV-positive blood donors).[57] The earlier decline in HIV infections probably can be traced in part to the very behavior modifications that would be undermined if Duesberg's controversial theory were ever to gain acceptance.

So, what are we to make of Duesberg and his theory? Has he (a) put at risk the lives of many people who may believe in his ideas, (b) sought publicity for his ideas without subjecting them to a serious critical analysis, or (c) served a useful role initially by forcing the community of AIDS researchers to demonstrate that they had not colluded in forging a premature consensus as to the nature of the disease? Probably, all of the above.

My rating for the idea that HIV is not the cause of AIDS is 3 cuckoos.

4 Sun Exposure Is Beneficial

In 1903, Niels Ryberg Finsen received the Nobel Prize in Medicine for demonstrating the therapeutic powers of sunshine in healing tuberculosis, and for identifying ultraviolet (UV) radiation as the responsible component. Following that discovery, the beneficial effects of sunshine on human health were widely accepted by both physicians and the general public. Enhanced sunlight exposure was sought by some patients in specially constructed solariums in order to treat a variety of ailments. Even today some health resorts offer such heliotherapy. Nowadays, however, we are more accustomed to hearing about the "dark side" of sunshine, most importantly with regard to skin cancer. Media stories of increased UV exposure due to a thinning ozone layer only add to these public concerns. Many sun worshippers, concerned more with their appearance than their health, use sunscreen in order to limit their exposure and avoid the discomfort if not the adverse health effects associated with sunburn.

Here we consider the controversial idea that sunbathing is on balance good for you, or that the harmful effects are more than outweighed by the beneficial ones. We want to consider whether the harmful effects may have been overdramatized and the beneficial effects greatly underappreciated by the general public and the medical community. Let's start with the harmful effects, since they are much better known.

Harmful Effects of Sun Exposure

Although sun exposure is a contributing factor to a variety of health problems, including cataracts and retinosis, skin aging, sunburn, and conceivably non-Hodgkins lymphoma,[1] the most serious worry surrounding sun exposure is skin cancer. Sun exposure is generally acknowledged as the main cause of skin cancer by reputable medical organizations, such as the International Association for Research on Cancer.[2] However, some medical societies have been fairly cautious in their assessment of the relationship between skin cancer and sun exposure. For example, in discussing malignant melanoma, the most serious form of skin cancer, which is responsible for about three-fourths of skin cancer deaths in the United States, the Family Medical Guide of the American Medical Association notes that "many years of exposure to strong sunlight *seem to play a part* in the development of the disease"(emphasis added).[3] This note of caution in assessing the sunshine–skin cancer link seems to be warranted in view of the "weak and conflicting evidence" many researchers find between melanoma and sun exposure.[4]

Elwood and Jopson, a pair of epidemiologists studying the melanoma–sun exposure link, have looked at twenty-nine separate studies done by researchers in a variety of countries.[4] These studies use the case-control method to compare the incidence of melanoma among people who received very high sun exposure on an intermittent basis and a control group who received normal or low exposures. Heavy intermittent exposure, such as you would get while swimming (and which would more likely lead to sunburn), is believed to be more important than continuous or chronic sun exposure as a risk factor for melanoma.[4] But the stud-

59 ies also looked at occupationally related sun exposures as well, which are more likely to be of the long-term continuous type. Altogether there were nearly 7,000 melanoma cases in the twenty-nine studies, which were of greatly varying size: the largest study considered 1,091 melanoma cases, while the smallest had only 58.

What Elwood and Jopson found in their analysis is rather strange. For those studies focusing on *intermittent* sun exposure there was a statistically significant increase in risk of melanoma when comparing people who had heavy exposures to a control group. The "relative risk" (RR) of the heavily exposed group was 1.71 ± 0.18, meaning that the heavily exposed group had a 71 ± 18% greater chance of getting melanoma than the controls. The quoted uncertainty range represents a *two*-standard-deviation limit,[1] and has been written here in terms of a symmetric range even though Elwood and Jopson quote slightly different upper and lower uncertainties.[5]

Surprisingly, for the studies involving occupational sun exposures, the average relative risk was less than 1.00 (meaning that sun exposure was protective against melanoma). The average of twenty studies yielded a relative risk of 0.86 ± 0.09 (2 std dev),[6] meaning that occupationally exposed people had 14 ± 9% *fewer* melanomas than controls (who were matched for other risk factors such as age and skin complexion). Why should intermittent sun exposure lead to more skin cancer, but occupational heavy exposure lead to less?

One plausible idea suggested by Elwood and Jopson is that occupational sun exposures tend to be of the long-term continuous type, and these lead to a protective mechanism

[1] A two-standard-deviation limit means that there is a 95 percent chance that the true value lies in the quoted range.

 of skin darkening and thickening. This hypothesis is supported by a study in which Elwood and coauthors looked at how melanoma risk varied with the estimated hours of sun exposure a person received per year. They found that for sun exposures in three ranges below 200 hours per year there was a significant increase in risk of about 50 percent (compared to those receiving less than 50 hours), but for higher exposures, the risk was the same as for those receiving under 50 hours.[7]

In twelve of the studies Elwood and Jopson looked at, researchers reported a relative risk associated with total sun exposure (including both intermittent and occupational exposure): RR = 1.18 ± 0.16 (2 std dev). In other words, based on total sun exposure, the expected increased chance of melanoma for heavily exposed people would be 18 ± 16%. As expected, this result is intermediate between the large increase in relative risk associated with intermittent exposures and the small decrease associated with occupational exposures.

Although the result is (barely) statistically significant at the two standard deviation level, there are reasons to regard it with some caution, primarily because the various studies are quite heterogeneous or inconsistent. For example, out of the twelve studies on total exposure, the chi square (a measure of inconsistency) is 62.1, which would happen by chance less than one time in a thousand, if the studies in fact are measuring the same quantity. As one indication of the inconsistency of the twelve studies, we note that four of them yield relative risks at least two standard deviations above one (sun exposure is harmful), while two yield relative risks at least two standard deviations below one (sun exposure is beneficial).

How should we report a single average value when reporting on many studies, some of which are inconsistent?

 One approach used by Elwood and Jopson is to drop from the average those studies which are the most severe outliers (i.e., that contribute the most to chi square). But that approach can possibly penalize studies that have the greatest statistical power, and that have accounted for some factor that everyone else missed. It also assumes that the studies are in fact all measuring the same quantity, which might not be the case here, since the relative risk of melanoma due to sun exposure might well be correlated with nationality and skin complexion. Another approach to reporting the average of many studies when some are inconsistent is to enlarge the quoted uncertainty by a scale factor that reflects the degree of inconsistency. It can be shown that a scale factor defined by the square root of chi square over one less than the number of studies is the appropriate choice.[8] In the present case, this yields a scale factor of $\sqrt{62.1/11} = 2.38$. Enlarging the uncertainty by a factor of 2.38 would mean that the risk of melanoma among heavily exposed people is $18 \pm 38\%$ greater than controls. Thus, in summary, the risk of melanoma based on total (intermittent plus occupational) sun exposure is of questionable statistical significance.

Based on the preceding discussion, it seems that sun exposure has a negative or positive effect on the risk of melanoma depending on whether the exposure is intermittent (especially if it leads to sunburn), or if it is occupational (indicating a long-term and continuous exposure). Thus, we may speculate that those sun bathers who use a lot of sunscreen, and who limit their sun exposure initially, may be on to something, not only for the purpose of getting a better tan, but also for reducing their risk of deadly skin cancer. Conceivably, those sunbathers who take such precautions may have more in common with the occupationally sun-exposed rather than those who get heavy intermittent exposures—in which case their risk would actually be below that of people who avoid the sun.

But in looking at the trade-off between the harmful and beneficial effects of sun exposure, let's avoid such speculation and concede conservatively that there may be some increase in the risk of deadly melanoma, the precise amount of which is unknown. The question then becomes: Does the benefit of sun exposure outweigh the possible harm?

Benefits of Sun Exposure

There are many well-established benefits linked with sun exposure, including therapeutic effects associated with tuberculosis, the skin conditions of dermatitis, psoriasis, and dandruff, the childhood disease rickets, osteomalacia (softening of the bones), some psychological disorders (SAD: seasonal affective disorder), and possibly even multiple sclerosis.[9] But most significant is the possible beneficial effect of sunshine in reducing the risk of coronary heart disease (CHD), the leading killer of Americans.

Coronary heart disease, which is the progressive blockage of the coronary arteries, is the most common form of heart disease. It is interesting to note that the death rate from CHD in the United States is approximately one hundred times that due to skin cancer, although any given individual might have a greater or lesser risk of the two diseases. As a result, even in the highly unlikely event that heavy sun exposure were found to double the mortality due to skin cancer, that harm would be offset by a mere one percent decrease in CHD mortality due to the exposure—at least for the U.S. population as a whole.

63 Under the assumptions just stated, there would be both winners and losers if sun exposure were to increase. The losers would be those who had a higher than normal risk of skin cancer and a lower than normal risk of CHD. The equally numerous winners would be those whose relative risk of the two diseases was reversed. If the risk of coronary heart disease mortality were decreased by *more* than one percent at the same time the risk of skin cancer mortality doubled as a result of increased sun exposure, there would be more winners than losers if sun exposure were to increase—though any given individual might be among the losers. In fact, some studies have shown that it is possible that the protective effect of sunshine in reducing CHD mortality could be far in excess of that hypothetical one percent "break-even" point.

Evidence for the protective effect of sunshine on coronary heart disease starts with the interesting observation that CHD death rates in various countries and regions seems to be strongly correlated with latitude, which is a rough measure of the amount of sunshine averaged over the year. For example, in one study comparing the CHD death rates in Belfast, Northern Ireland, and Toulouse, France, it was found that the CHD death rate among men aged 55–64 was a remarkable 4.3 times higher in Belfast.[10] It is particularly interesting to note that when all the nondietary risk factors were compared for the two populations, including smoking, obesity, high blood pressure, etc., there were almost identical overall risks of CHD between them. Moreover, even the diets in the two locations were remarkably similar, although there was more cholesterol and red wine intake in Toulouse and more saturated fat intake in Belfast.[10] Extremely high CHD mortality is common not only in Belfast, but in northwest Great Britain generally, which is also a region of little annual sun exposure due to both the extensive cloud cover and the high latitude.[11]

Higher latitudes (greater distance from the equator) are at locations where the sun spends more time low in the sky, leading to less integrated sun exposure during the course of a year. Ultraviolet radiation is even more affected by latitude than visible radiation, because the amount of UV reaching the Earth's surface depends on the thickness of the ozone layer at a given latitude and also the angle of the sun. When the sun is low in the sky, the sun's rays pass obliquely through the absorbing ozone layer, and the amount of UV reaching the surface is much reduced. Consequently, the amount of UV exposure you receive is greatest at midday, during the summer, and at low latitudes.

But the observed link between latitude and CHD mortality by itself is merely suggestive of a relationship between sun exposure and heart disease. To make a convincing case for a relationship we need to look at the many ways such a connection might show itself, see whether confounding variables might also explain the observed link, and finally determine whether a plausible biological mechanism exists for the link. Let's start with the last issue.

Grimes, Hindle, and Dyer have investigated the relationship between sun exposure and CHD and claimed that sun exposure is protective against heart disease.[12] They believe that the underlying mechanism is based on the production of vitamin D in a person's skin as a result of sun exposure. The theory is that since cholesterol and vitamin D are both made from the same precursor compound (squalene), then if your body makes vitamin D it uses up the squalene, leaving less to make cholesterol—a well-known risk factor for CHD. The suggested protective effect of vitamin D, and its inverse correlation with cholesterol, is supported by a study which found that heart attack victims'

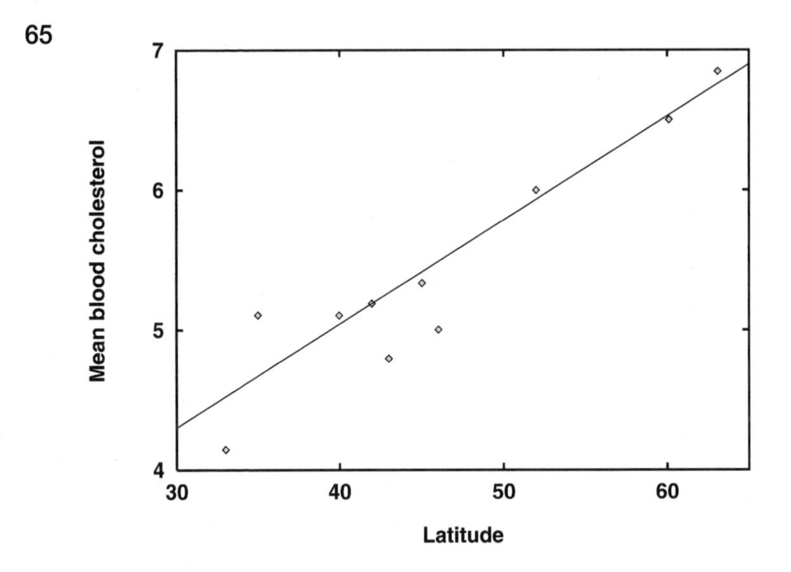

Figure 4.1. Average blood cholesterol level in mmol/l versus latitude, according to Grimes, Hindle, and Dyer (see endnote [12]).

blood had statistically significant lower levels of vitamin D, as well as HDL (the "good" cholesterol), but higher levels of total cholesterol.[13] This finding, however, is only suggestive of a relationship between cholesterol and vitamin D. It is possible that the lower levels of vitamin D found in heart attack victims occurred because they had been feeling poorly and spent less time outdoors in the weeks before the attack than the control group. In that case, their low vitamin D levels would be due to their heart disease, not the reverse.

In their attempt to link sun exposure to CHD, Grimes, Hindle, and Dyer looked at the connection between mean blood cholesterol for people living in different cities and countries. They found a significant connection with latitude, which would occur on a chance basis less often than 0.1 percent of the time, i.e., $p < 0.001$) (see fig. 4.1). This connection is to be expected, given their hypothesis: lower

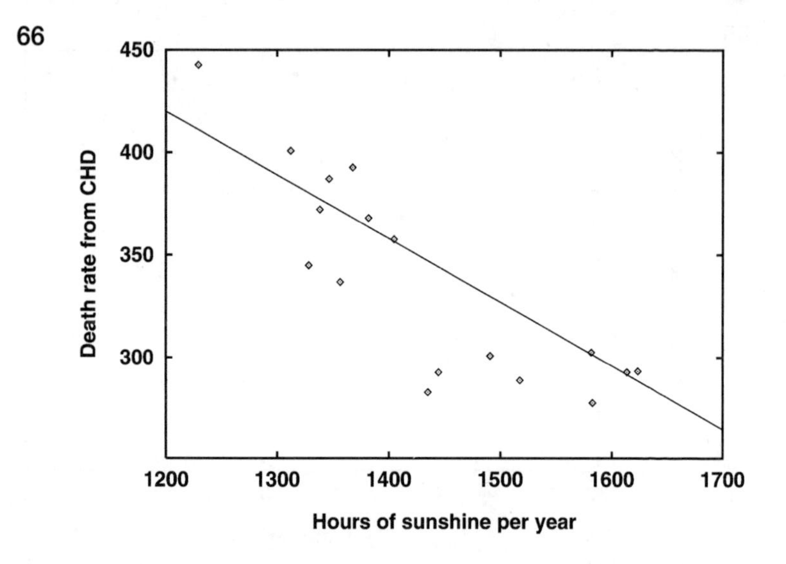

Figure 4.2. Average death rate from coronary heart disease per 100,000 males versus hours of sunshine per year according to Grimes, Hindle, and Dyer (see endnote [12]).

latitude implies more sun exposure, which implies more vitamin D production, which implies less cholesterol production.[12] Grimes and coauthors also found a significant variation ($p < 0.001$) in CHD deaths with hours of sunshine per year for people living in different parts of Great Britain whose amount of sunshine per year depended much more on distance from the Atlantic sea coast than on latitude (see fig. 4.2).[12] (Note that hours of sunshine are a good indicator of annual UV exposure for people living in similar latitudes, but not in general.)[2]

This observed correlation between CHD death rate and annual sunshine exposure is helpful in ruling out confound-

[2] If we compare people living in different latitudes, the annual UV exposure also depends strongly on latitude, which determines the angle of the sun above the horizon.

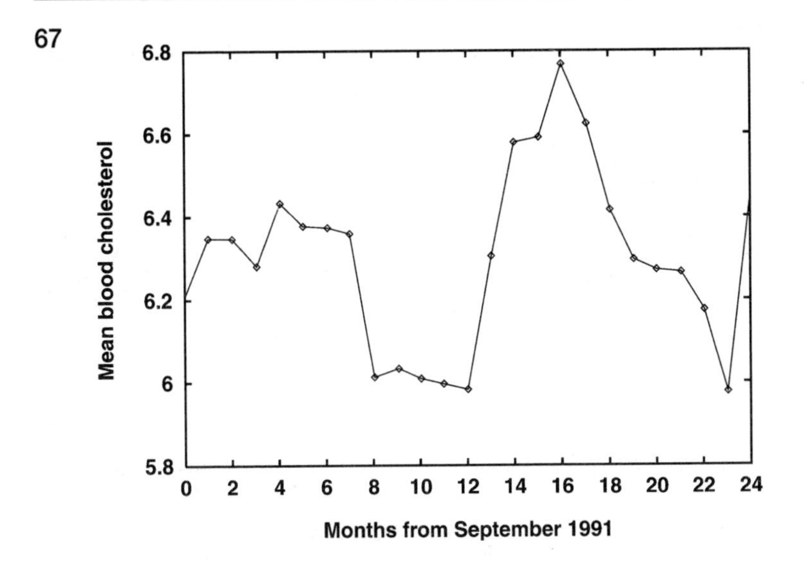

Figure 4.3. Average cholesterol level (in millimoles per liter) among residents of Blackburn, Lancashire, each month during the two-year interval September 1991–September 1993, from reference in endnote [12].

ing variables, because people living in different parts of Great Britain have less nonuniformity of diet and ethnic heritage than people in different countries. Grimes and coauthors have also looked at the mean cholesterol levels found during the course of a two-year period for residents in one British city (see fig. 4.3).[12] The location of the peaks and valleys of this graph appear to correlate inversely with the peaks and valleys of the hours of sunshine during each month (see fig. 4.4). Even the differences between the two succeeding winters agree: higher cholesterol levels and less sunshine in the second winter than the first.

Figures 4.3 and 4.4 strengthen the case for a connection between sunshine and CHD, because they involve a relatively uniform population. Could changes in diet during the course of the year cause the seasonal variation in cholesterol

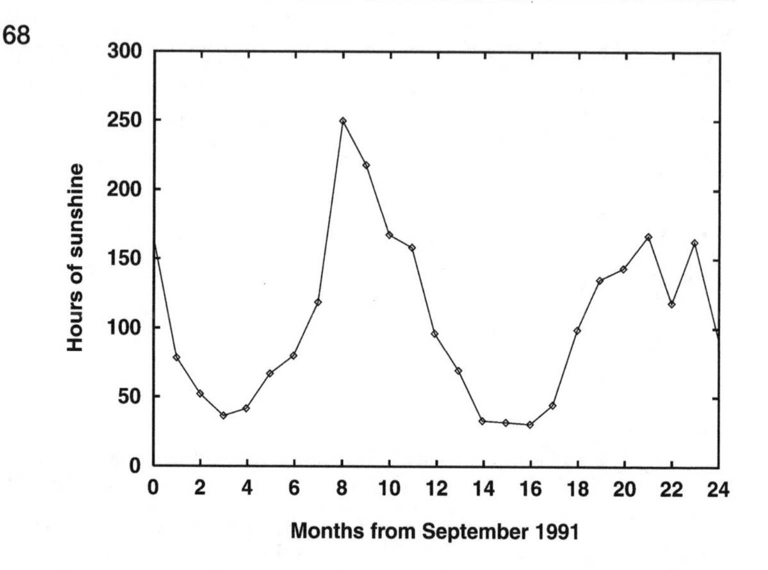

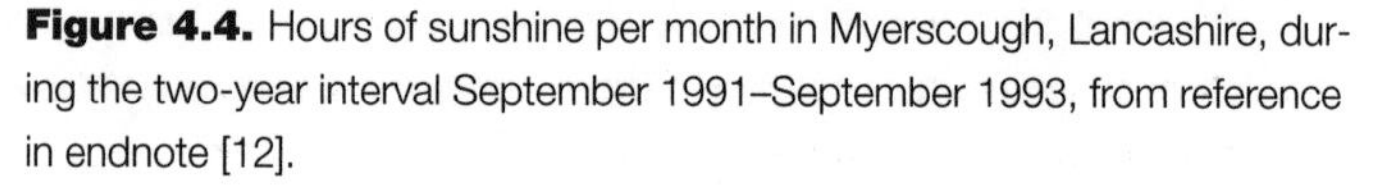
Figure 4.4. Hours of sunshine per month in Myerscough, Lancashire, during the two-year interval September 1991–September 1993, from reference in endnote [12].

levels? Grimes and coauthors have investigated this possibility, and they believe the answer to be negative. They looked at thirteen different food categories (including meat, fish, fats, eggs, and cheese) and found that the greatest variation between January through March and April through June was at the 10 percent level for any one food category. In the "suspicious" category of fats, there was zero seasonal change, and in the eggs category there was 9 percent *greater* consumption in April–June (when cholesterol levels tended to drop).

Another plausible variable besides diet that could easily be confounded with annual hours of sunshine is regional temperature. Sunshine and temperature are closely correlated, both with respect to their geographical and temporal variation. Some researchers have found, for example, that blood pressure (another risk factor for CHD) correlates with temperature. So, how can we be sure that the correlations seen in figures 4.1 through 4.4 show that sunshine rather than warmer temperatures is not the real cause of lower CHD mortality in warmer versus colder regions and during summer versus winter months?

One way that Grimes and coauthors have shown that sun exposure is the relevant variable rather than temperature or diet is by looking at how cholesterol levels depend on garden ownership (a proxy for higher than average outdoor sun exposure during the summer). They found that the eighty-one garden owners had statistically significant higher levels of vitamin D ($p < 0.025$) and lower levels of cholesterol ($p < 0.01$) than the sixty-one non-garden owners, but *only* during the summer months, as would be expected.

A second reason for believing that sun exposure rather than temperature is the key variable is supplied by U.S. studies that show that CHD mortality decreases with altitude.[14, 15, 16] Higher altitudes generally have greater sun exposures but lower temperatures, and altitude, therefore, would be protective if sun exposure were the relevant variable, but harmful if warm temperature were responsible for lower CHD mortality.

A third reason for doubting that changes in temperature rather than sun exposure are responsible for variations in CHD mortality is provided by a study that looked at seasonal mortality rates in Los Angeles and New York.[17] The summer-winter temperature variation between the two cities is far greater than any difference in the amount of sunshine during the year. (Angelenos' familiarity with snow is probably comparable to New Yorkers familiarity with hominy grits.) In fact, the average summer-winter temperature difference in New York is 2.5 times greater than that in Los

 Angeles, even though the summer-winter mortality difference is nearly the same (around 20 percent) in the two cities.[17] This fact indicates that summer-winter CHD mortality differences are probably not linked to temperature changes.

According to Grimes and coauthors, diet is also unlikely to be a confounding variable. Diet may play some role in risk of CHD, but it appears to be far less significant than other known risk factors, such as smoking, high blood pressure, obesity, family history, or a sedentary lifestyle. In fact, dietary manipulation as part of trials to prevent CHD has been very unsuccessful, and overviews of such trials show no overall benefit.[18, 19] Therefore, it seems unlikely that diet is a confounding variable that could explain the different CHD rates seen in different countries, different parts of Great Britain, and different seasons.

We can also establish whether the key variable in determining CHD mortality is geographically based (such as sunshine abundance) by considering studies of people who migrate from one country to another. Such studies show that immigrants tend to have the relative risk of CHD associated with their adopted country, even though they tend to continue their previous diet.[20] A particularly instructive example are Asian immigrants to Great Britain, who tend to have very high rates of CHD.[21, 22] These immigrants usually have little leisure time to take holidays in the sun, lead a mainly indoor life, and tend to wear clothes leaving little skin exposed. In their original tropical climate they may get an adequate sun exposure despite these habits, but perhaps not in Great Britain. An indication that the accumulated deficit of sun exposure received by Asian immigrants is unhealthful after several years (when their vitamin D reserves have been depleted) would be their higher than normal incidence of various diseases associated with inadequate sun exposure: rickets, tuberculosis, and osteomalacia.[23, 24, 25]

In summary, it would seem that a plausible but not yet convincing case has been made for the idea that sun exposure is protective in reducing coronary heart disease. The case would be considerably strengthened if case-control studies were done (of the type studying the link between sun exposure and skin cancer), and if specific individuals receiving varying estimated amounts of sun exposure were found to have both increased vitamin D levels and reduced cholesterol levels following such exposure.

My rating for the idea that sun exposure is beneficial is zero cuckoos.

What About the Ozone Layer?

At ground level, ozone is a toxic pollutant produced in automobile exhaust that contributes to respiratory problems. Severe ozone buildups sometimes lead local governments to issue air-quality alerts. Although it may be harmful at ground level, ozone high up in the stratosphere plays a crucial role in shielding the planet from deadly levels of UV solar radiation. Without any ozone layer, some have suggested life on Earth would be cooked—literally.

Beginning in the 1970s, scientists began to suspect that chlorofluorocarbons (CFCs), widely used as refrigerants, were finding their way into the stratosphere and were responsible for a depletion of the ozone layer. The extent of the depletion varied considerably with location. In an ozone "hole" over Antarctica, levels were found to drop by 50 to 70 percent during springtime, covering a region across 10 percent of the Southern Hemisphere. In other places, the levels of ozone depletion have been of more modest proportions: a drop of 8 percent over the U.S. by the early 1990s.

CFCs have been banned by the 1987 Ozone Treaty and replaced by more ozone-friendly chemicals, but it is predicted that the ozone layer will take decades to recover because chlorine molecules have a long residence time in the atmosphere. Was the international community correct in heeding the warnings regarding a depleted ozone layer and banning CFCs before all the evidence was in? If sunshine (and UV exposure in particular) is really on balance beneficial, should we now reverse that decision? Apart from the political impossibility of such a move, our state of knowledge on the benefits versus risks associated with sun exposure certainly would make such a decision highly premature. But it is interesting to ponder the question of "winners and losers." Enhanced UV exposure associated with a thinning ozone layer might well be of benefit to humans in regions of the globe now receiving very little sun exposure, and might well be harmful to people elsewhere.[26] One cannot help but wonder whether an international treaty banning ozone-destroying CFCs would have been so easily agreed upon had this winners and losers aspect been better appreciated.

5 Low Doses of Nuclear Radiation Are Beneficial

"HORMESIS" is a word not found in many dictionaries. That's unfortunate, because the concept deserves to be better known. It refers to the notion that large doses of an agent are harmful but very small doses are beneficial. Hormetic effects have been demonstrated in humans for a variety of agents including alcohol, sunshine, iodine, copper, sodium, potassium, and even cholesterol. Many vitamins (such as A, D, and B_6) and minerals (such as calcium and iron) often taken as dietary supplements to improve health also can be harmful if taken in large doses.[1] Even small doses of harmful bacteria can be beneficial when they stimulate the body's immune system. It is well known that people living in an overly sterile environment or one in which particular bacteria are not present are particularly susceptible if they are later exposed to the organisms. From the length of the list of examples of hormetic agents, it might seem as if hormesis is the rule rather than the exception. Here we shall consider the controversial question of whether nuclear radiation might also be hormetic.

Nuclear radiation (also called ionizing radiation because it creates a trail of ions in air or other media) is emitted by radioactive nuclear isotopes. It comes in various types, including alpha particles, beta particles, neutrons, and gamma rays, which are a very high energy type of electromagnetic radiation. Other kinds of electromagnetic radiation, such as ultraviolet and microwaves, are not ionizing, even though they are harmful to humans in large doses.

 Often, in referring to nuclear radiation throughout this chapter, we may leave off the "nuclear," which can cause some confusion, because in another context, the term radiation can also refer to all types of electromagnetic radiation, even including visible light.

What are the biological effects of radiation on humans? Very large doses of radiation cause radiation sickness, which is often followed by a very painful death usually within a month of the time of exposure. Doses insufficient to cause short-term effects have also been shown to result in serious problems many years after a radiation exposure, including increased susceptibility to dying from cancer. But what about low doses? Is it possible that nuclear radiation might actually be beneficial when received in doses comparable to those received as part of the natural background radiation? Although the idea of improving one's well-being through the nuclear version of tanning salons may sound patently absurd, the idea of nuclear radiation hormesis should not be dismissed out of hand.

We are continually being exposed to nuclear radiation in the natural environment. About half of the average person's radiation exposure is due to radon gas seeping up from the ground, and most of the rest is due to cosmic rays from space. The amount of uranium ore in the soil varies from place to place on Earth. Since uranium is the primary source of radon, these variations lead to variations in the amount of radon gas seeping up the ground. Your exposure to cosmic rays depends primarily on altitude, since this radiation from space gets more reduced, the thicker the atmospheric blanket through which it passes. As a result, astronauts are subject to appreciably higher radiation levels in space, and even people riding in jet aircraft receive additional radiation exposures during their flight. However, the additional exposure is a very small fraction of your annual total dose, since total time in the air is a tiny fraction of your lifetime—although if you are a frequent flyer, it may seem otherwise.

The conventional wisdom about nuclear radiation is that the biological harm done is linearly proportional to the dose received. This concept is known as the linear-no-threshold (LNT) hypothesis. According to LNT, for example, your increased risk of dying from cancer as a result of exposure to background radiation would double if you lived your whole life in an area where the level was $2x$ rather than x. This conclusion holds no matter how small x is. (Note that we are referring to *increased* risk levels above some spontaneously occurring level that is not caused by radiation, so we are certainly not saying that your risk itself doubles if the background radiation doubles, only the increase—which is usually a very tiny fraction of the spontaneously occurring level.) For example, let's assume that a sudden radiation dose of 250 cSv would roughly increase your chances of dying from cancer by 100 percent. (Radiation units will be defined later.) If LNT is true, then we can conclude that a dose of 2.5 cSv would increase your chances of dying by one percent, and the 0.025 cSv dose from a single chest X-ray increases your chances of dying from cancer by 0.01 percent.

The LNT hypothesis is the basis for the commonly heard saying that there is no "safe" level of radiation below which no harm will occur, or that any radiation exposure carries some risk. ("Safe" is being used here in the same sense as saying that no place on Earth is safe from the threat of an asteroid impact—the threat may be very small, but it is never zero.) LNT is, of course, in direct conflict with the notion of hormesis, which claims that not only is there a safe level of radiation, but that you are actually improving your health (or decreasing your risk of cancer) by being exposed to radiation up to some level. There are, of course, still other possibilities besides hormesis and LNT, the most obvious being that a threshold exists below which radiation is neither harmful nor beneficial.

Why should it be difficult to sort out which of the three possibilities (hormesis, LNT, or a threshold) best describes the actual effects of nuclear radiation on humans? The primary difficulty in sorting this out is threefold: (1) the number of spontaneously occurring cancers and other diseases is far greater than the number attributable to radiation; (2) people are subject to a variety of harmful agents in their environment, only some of which are known; and (3) cancers due to radiation are indistinguishable from cancers due to other environmental causes.

The quantitative relationship between the radiation dose received and the harm it produces is usually called the *dose-response function*. Here we shall take a look at different types of studies to see if one of the three types of dose-response functions mentioned previously (hormetic, LNT, or threshold) (see fig. 5.1) is in fact favored over the others based on the available data. (As already noted, it is possible that the correct dose response function is none of these three types.) The data we will look at come from a variety of sources, including studies of (1) survivors of the Hiroshima and Nagasaki bombings, (2) workers in the nuclear industry, and (3) people living in areas of high and low background radiation.

Japanese A-Bombings, and Radiation Doses and Units

Of the 429,000 people believed to have been living in Hiroshima and Nagasaki at the time of the U.S. bombings in 1945, approximately 170,000 perished due to the bombings by the end of that year (100,000 in Hiroshima and 70,000

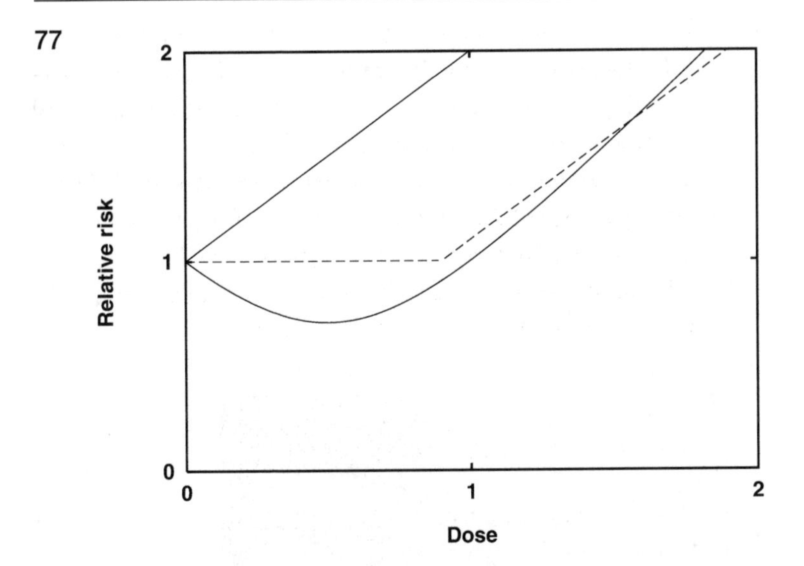

Figure 5.1. Hormetic, threshold, and linear-no-threshold (LNT) dose-response functions. Below a dose of 1.0 (in arbitrary units), the hormetic function has a relative risk less than 1.0 and is, therefore, protective. In this illustration, the threshold dose-response function is shown by the dotted line.

in Nagasaki).[2] The survivors of these bombings (who are known in Japan as *hibakusha*) have felt the aftereffects throughout their lives, and they have been regarded by other Japanese as being in a real sense "contaminated," and unfit for marriage, for example. The health of the hibakusha has been carefully monitored over the years by the Atomic Bomb Casualty Commission (1950–74) and the Radiation Effects Research Foundation (1975–). Although the deaths of nearly 200,000 persons, including many women and children, was a great tragedy, it does not diminish that tragedy to learn from its survivors as much as we can about the effects of radiation on humans.

The number of deaths among atomic bomb survivors between 1950 and 1985 is 20,777.[2] By comparing the mortality rates of the survivors with those of Japanese living in or near Hiroshima and Nagasaki who were not exposed to radiation, scientists are able to determine the relative risk of mortality for people receiving various radiation doses. By definition, the relative risk is simply the ratio of the observed number of deaths to the expected number of deaths among unexposed individuals. For example, a relative risk of 1.00 means that exposed individuals have the same mortality as unexposed individuals, and a relative risk of 1.50 means that their mortality is 50 percent higher.

Figure 5.2. The shadow of a man standing next to a ladder, burned into a wall in Nagasaki by the intense flash of thermal (non-nuclear) radiation. (United Nations Photo)

The radiation dose received by individual Japanese survivors can be roughly estimated based on where they were at the instant the bombs were detonated, specifically taking into account their distance to ground zero and any shielding from the radiation due to buildings or terrain. The units for measuring radiation dose are a bit confusing and will require some discussion. A radiation dose unit known as the Gray represents 1 Joule of energy deposited in each kilogram of body mass, as the radiation passes through the body.[1] (Often, we will use the "centiGray" [one-hundreth of a Gray], abbreviated cGy, which corresponds to an older unit known as the rad.)

Japanese survivors of the bombings received doses ranging from 0 to over 450 cGy—the dose usually regarded to be lethal to half the people exposed. How do such doses compare with those received during peacetime? In the course of a year the average person receives a dose from background radiation of about 0.3 cGy plus another 0.02 cGy for each medical X-ray he or she receives. A typical lifetime dose from background and medical exposures might be around 25 cGy spread over seventy years. We have ignored radiation doses from other causes here such as nuclear power plants, which in their normal operation contribute less than 0.3 percent of the average person's radiation exposure.

What about nuclear reactor radiation releases when they are not operating normally? Surprisingly, even in the case of the disaster at Chernobyl, the worst nuclear accident in history, the radiation dose received each year by someone living in the most heavily contaminated areas surrounding Chernobyl would be only about 3 cGy. Thus, the radiation exposures received by some Japanese survivors far exceeds that received by those who had been living close to Cherno-

[1] One form of nuclear radiation known as neutrinos passes through our bodies in much greater abundance than any other, but since neutrinos hardly ever interact with other matter, they deposit almost zero energy inside us, and cause essentially no harm.

 byl, and of course also greatly exceeds exposures from normal background radiation. (Of course, there were also many Japanese living in Hiroshima and Nagasaki at the time of the bombings who received very low radiation doses, based on their location.)

The comparison between radiation doses received by Japanese survivors and doses received by individuals in normal peacetime or as a result of the Chernobyl disaster are somewhat misleading, however, in one important respect. Background radiation gives you a continuous radiation exposure over your entire life. In the case of Chernobyl, even though the radiation levels were most intense in the immediate aftermath, the fallout from the accident also creates a continuous exposure over many years to anyone not evacuated from the contaminated areas. The fallout occurred when the graphite core of the reactor burned, and radioactive particles were carried by the smoke high into the atmosphere and later came back down to Earth with precipitation. Unfortunately, this fallout cannot be cleaned up or decontaminated because the radioactive particles are spread over many thousands of square miles. Therefore, people had to be evacuated permanently from the most heavily contaminated areas.

Unlike at Chernobyl, little nuclear fallout was produced in the Japanese bombings. The A-bombs created fireballs that did not touch the ground and lead to fallout. As a result, the Japanese survivors received a radiation dose at the instant of detonation rather than over a period of many years. This difference made the radiation doses received by the Japanese survivors even more potent on a Gray-for-Gray basis, because radiation doses received over short times are much more harmful than the same dose spread over a long time—possibly because the body is able to repair the damage when the dose is spread out over time.

A further complication is that different types of radiation emitted by radioactive nuclei cause different levels of harm, even if the number of Grays is the same. Thus, for example, a dose of one Gray from neutrons is much more harmful than one Gray from gamma rays. Gamma rays deposit their energy in a uniform manner all along the path of the gamma ray as it passes through the body and leaves a trail of positive and negative ions in its wake. In contrast, neutrons cause no trail of ionization, and instead deposit energy only when they collide with atomic nuclei. As a result, the energy deposited by neutrons is greatly concentrated rather than linearly spread along their path through the body.

Consequently, different forms of radiation are said to have different "relative biological effectiveness" (RBE). The unit known as the Sievert is often used in order to take into account the RBE caused by equal numbers of Grays of different types of nuclear radiation. By definition, one Sievert of any type of radiation is equally harmful. Thus, a one-Sievert dose of gamma rays is equivalent to one Gray of gamma rays, but one Sievert of neutrons is equivalent to perhaps only 0.1 Gray of neutrons, because neutrons have an RBE about ten times that of gamma rays. We often will use the unit the "centiSievert" (one-hundreth of a Sievert), abbreviated cSv, which corresponds to an older unit known as the rem.

Studies of Hiroshima and Nagasaki Survivors

When we compare mortality rates of Japanese A-bomb survivors with those of other Japanese, we find that the survivors actually live longer.[3] Could this finding be evidence for radiation hormesis? Not according to those researchers conducting the survivor studies who attribute the finding to the "healthy survivor effect." The idea is that those individuals who managed to survive the bombings and the horrendous conditions in their aftermath had to come from healthier than average stock. Longer lifespans among survivors *could* be evidence of hormesis—the beneficial effect of a low dose of radiation—but only if it were shown that those individuals who received a low dose of radiation lived longer than those who experienced either zero radiation or moderate to high doses. But, in fact no such result is found, unless we are highly selective in the way we examine the survivor data—essentially forcing the data to show the very effect we are seeking. Let's see how this data manipulation technique works in a specific example.

For example, consider the deaths during 1970–88 classified by sex, cause of death, and dose received for Nagasaki survivors (see fig. 5.3 for males only).[4] Statistically significant values of the relative risk below 1.00 imply that the number of deaths for a particular dose were below the expected number. Sohei Kondo, a proponent of hormesis, claims that the data point for the dose interval 50–99 cGy shows a statistically significant reduction in mortality from noncancerous diseases for men—the low point on the solid curve in figure 5.3.[2] The point certainly lies below a relative risk of 1.00, but is it really statistically significant?

According to Kondo, the deviation of this one point (19.5 deaths observed vs. 30 expected) is slightly more than two standard deviations ($p < 0.05$) using a chi square test.[2] In some studies in the biomedical and social science fields, $p < 0.05$ is regrettably regarded as the criterion for a deviation being statistically significant. (By definition, if $p = 0.05$ then such deviations may be expected to occur for one out of every twenty data points on the basis of chance.)

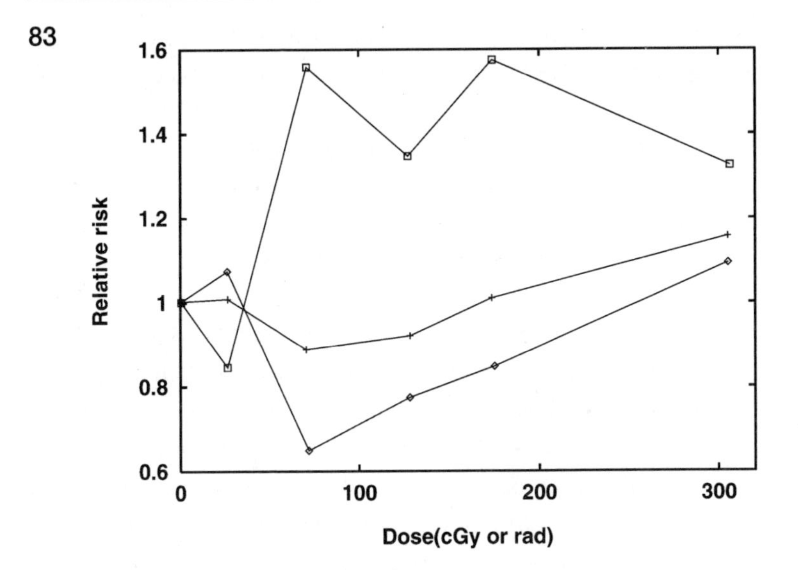

Figure 5.3. Relative risk of death for male survivors of the Nagasaki atomic bombing versus dose received: + for deaths from all causes, squares for deaths from cancer, and ◊ for deaths from causes other than cancer, from reference in endnote [2]. The data points actually represent relative risks for people receiving a *range* of doses. For example, the low point on the non-cancer mortality curve at the dose 75 cGy corresponds to a range of 50 to 99 cGy.

But the main statistical error here in claiming statistical significance is not in the choice of how many standard deviations one regards as being statistically significant, but rather in the use of arbitrary selections of a particular subsample of a total data set so as to enhance the very effect one is looking for.

For example, why just consider noncancerous disease mortality? Why just consider men? Why just consider deaths during the interval 1970–88? Why just consider Nagasaki and not Hiroshima survivors? The answer is that no such effect is seen unless one makes these particular choices. In fact, there is no reduction in the relative risk at a dose of 50–99 cGy (but rather an *increase*) when you look at the data for women, or the data for deaths due to cancer, or the deaths in years other than 1970–88.

It is very easy to see in this example how two standard deviation ($p < 0.05$) departures can occur routinely on the basis of chance. Suppose one believes that there will be a low data point in the low-dose region (say either 1–49 or 50–99 cGy). If there are two choices of cities to look at, two sexes, two categories of mortality (cancer and all other categories), two low-dose intervals, and two intervals of years of death (before and after 1970), that gives a total of five variables, each of which can take on two values, for a total number of 2^5 = thirty-two combinations of the five variables. It would be highly surprising if at least one or two of these thirty-two subsets of the data did not show a two standard deviation on the basis of chance.

The statistical error of choosing a subset of a total data set in order to enhance the statistical significance of one's claim is known as making an "informed choice" and also "cooking" the data. The making of informed choices is sometimes done without realizing it, and when done consciously, it is usually done with the purest of intentions, but it remains an exceedingly common way of inflating the statistical significance of one's results. It is an error to be carefully guarded against. (Very often, researchers making informed choices are able *retroactively* to justify why the particular effect they are seeking should be more prominent in the particular subset of the data they have selected. But retroactive justifications of informed choices are improper.)

Figure 5.4 shows the relative risk of cancer mortality for all A-bomb survivors (males and females) for the full period (1950–90).[5] As we can see, the data are quite consistent

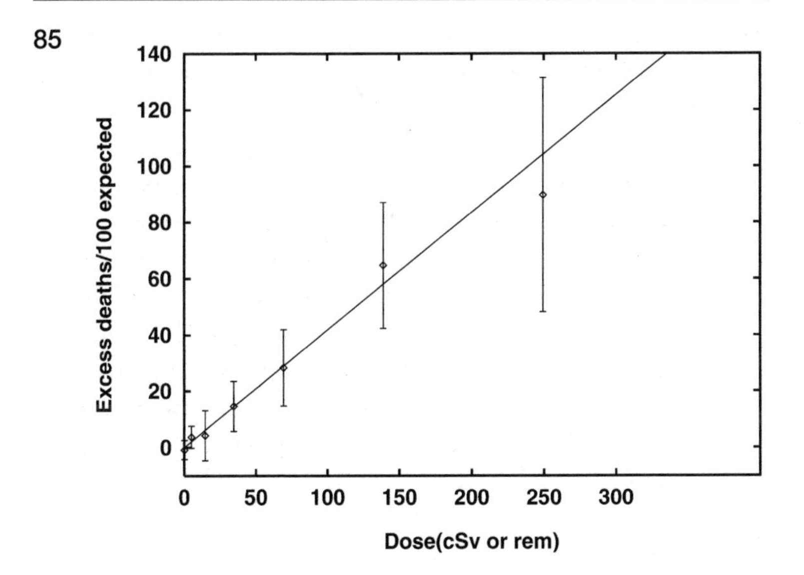

Figure 5.4. Excess deaths per one hundred expected for atomic bomb survivors versus radiation dose received in cSv during the period 1950–90, according to reference in endnote [5]. The error bars are two standard deviation limits.

with a linear dose-response function (the linear no-threshold model). Based on the slope of the best-fit straight line, we see that a survivor's chances of dying from cancer doubles at a dose of about 250 cSv, which is known as the "doubling dose." Despite the good fit to a straight line, we cannot rule out other dose-response relations in the low dose region.

Bernard Cohen, a critic of the LNT dose-response hypothesis, has noted that the data of figure 5.4 are also consistent when there is a threshold at a dose around 30 cSv, and maybe there is even a hormetic dose-response function at these low doses. Given the size of the error bars, Cohen's suggestion is certainly true. Of course, no matter how good the statistical quality of a data set, you can always postulate a threshold sufficiently close to zero that would be consistent with the data, assuming the data actually follow a linear relation.

Although the A-bomb survivor data give no statistically significant support to either the hormesis hypothesis or the idea of a threshold, neither do they refute these possibilities. Nevertheless, the survivor data can set limits on the size of these effects. For example, based on the size of the *two*-standard deviation error bars in figure 5.4, if there is any hormesis for low doses (say under 30 cSv), any reduction in relative risk below 1.00 would probably be no greater than about 5 percent, i.e., a curve dipping to about –5 in the figure.

Occupational and Epidemiological Studies

Before the hazards of radiation were well understood, workers in some occupations were exposed to extremely high doses. One such unfortunate group of female workers painted watch dials using radium to make watches glow in the dark. These women often ingested radium when they put the small brushes in their mouths in order to keep them pointed. The doses received by these workers were localized rather than whole-body doses, so that a number of them survived doses that on a whole-body basis would surely have been fatal. The data for these workers, shown in figure 5.5, are perhaps the most frequently cited evidence against an LNT (straight line) dose-response relationship, and for a threshold.[6] The curve drawn connecting the data points indeed reinforces the impression of a threshold—here much higher than any possible threshold in figure 5.4. The data points at doses below 1000 cGy lie on the *x*-axis, because no tumors were observed for those workers, but the

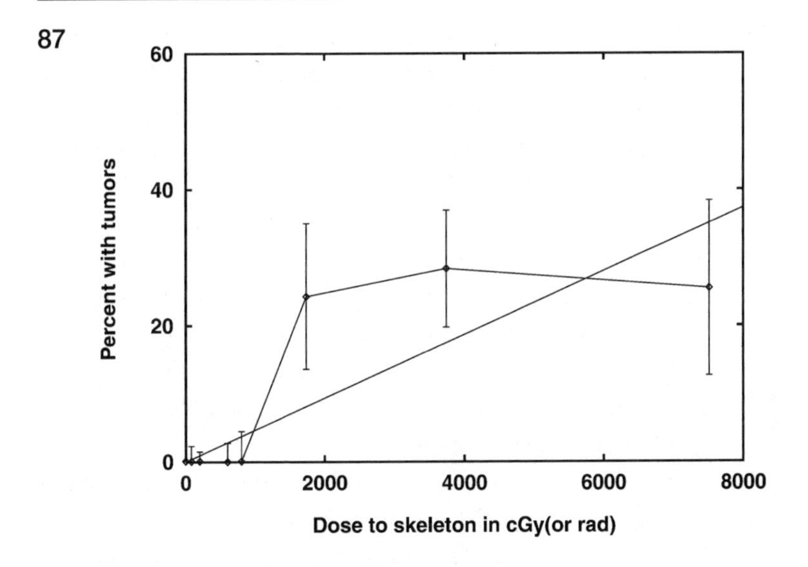

Figure 5.5. Percentage of workers who ingested radium who had tumors in the bone or head versus skeletal dose received in cGy, and the best-fit straight line, from reference in endnote [6].

tops of the error bars show what they would have been if one tumor had been found in each case.

Do these data really support the idea of a threshold and contradict LNT as claimed, for example, by Bernard Cohen?[7] The sloping straight line—a best fit to the data points shown in figure 5.5—has a chi square probability of 46 percent. Based on this good fit, we see that in fact the data fit LNT reasonably well. The "evidence" for a threshold comes primarily from the visual impression created by connecting the data points. Moreover, if we really were to take that curve literally, we would also have to accept that (1) the threshold here is perhaps thirty times higher than for the Japanese survivors, and (2) the harm done is essentially independent of dose once the dose exceeds the threshold, neither of which seems plausible.

 What kind of definitive study might confirm or refute whether or not a threshold or hormesis exists at low doses? Obviously, now that we are more aware of the dangers of radiation, it would be highly immoral (and in fact criminal) to do experiments on humans exposed to various levels of radiation in order to improve our knowledge and test whether or not a threshold (or hormesis) exists. However, since people are naturally exposed to different levels of background radiation, we can use epidemiological studies to test these hypotheses. Epidemiological studies could in principle involve the whole population, and therefore have great statistical significance.

In 1973, a large-scale study was reported by researchers at Argonne National Laboratory in Illinois which looked at the age-adjusted cancer mortality rates for people living in each state in the U.S.[8] The results are shown in figure 5.6, which is a plot of cancer mortality versus average background radiation for each of the states. Interestingly, the seven states (Idaho, Montana, Utah, New Mexico, South Dakota, Wyoming, and Colorado) with the highest level of background radiation (above 1600 μSv/year) all have the *lowest* cancer mortality rates. (The U.S. average cancer mortalitity during the study period was 150 deaths per 100,000.)

If the individual points in the figure were randomly located dots, the odds of having these seven points so far below the average would be less than one chance in 100,000.[8] The negative correlation between background radiation levels and cancer mortality shown in figure 5.6 would at first glance seem to offer good evidence for hormesis. In fact, if the results are taken at face value, and a straight line were fit to the data, the slope of that line would imply that, at low doses, radiation is ten to twenty times as beneficial as it is harmful at higher doses. But just how solid is the evidence for hormesis here?

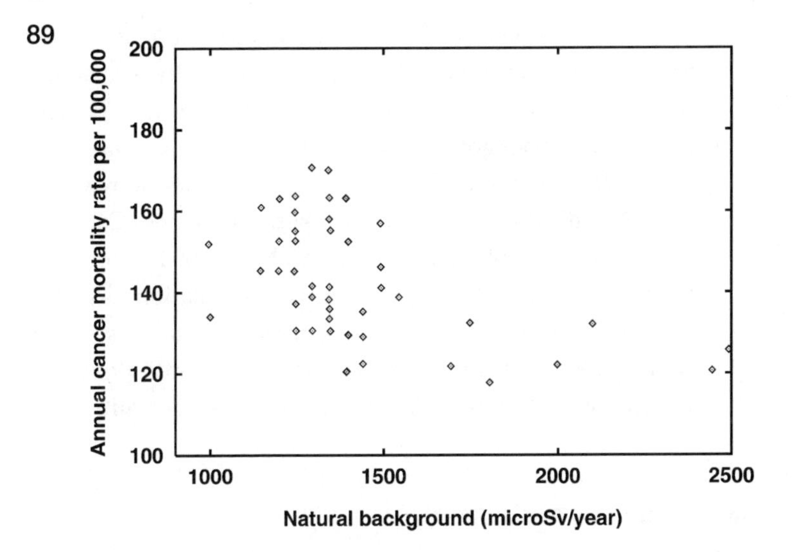

Figure 5.6. Cancer mortality per 100,000 persons in the United States by state between 1950 and 1967 versus natural background radiation in μSv/year in those states, according to reference in endnote [8]. The seven states with the highest background radiation are identified in the text. The graph shows forty-five states.

One problem with the data in figure 5.6 is that if the negative correlation were due to hormesis, the size of the effect is much too *great*. For example, the seven identified states have cancer mortality rates that are on the average about 15 percent below the national average. The annual dose to residents of these states is on the average about 2000 μSv, which is about 700 μSv higher than the U.S. average. Thus, people in those seven states would on average receive a 3.5 cSv higher dose than the U.S. mean over a 50-year period, i.e., 50 yr × 700 μSv/yr = 3.5 cSv. But there is no evidence from the Japanese survivor data that the dose-response curve in figure 5.4 could dip by as much as 0.15 (15 percent) for the very low dose region. (In fact, we earlier estimated a maximum dip of only 5 percent.)

Another problem with the data in figure 5.6 is that there are many environmental differences between the states, and some of these differences (such as altitude) are strongly correlated with background radiation levels. For example, the seven high-background states all have altitudes above 1,000 meters, with the two highest background states, Colorado and Wyoming, both almost entirely above 2,000 meters. The seven states also have much lower population densities than average. Both of these environmental factors would subject residents to much less air and water pollution, which might well explain their lower cancer mortality rates.

Bernard Cohen's Radon Study

Perhaps the greatest support for hormesis has been a widely quoted study by Bernard Cohen of the University of Pittsburgh.[7] This study involved the measurement of radon levels in hundreds of thousands of homes in 1,729 U. S. counties that contain 90 percent of the U.S. population. Cohen calculated the radon levels averaged over each county and looked at the correlation with the average lung cancer mortality in those counties.

Many epidemiological studies besides Cohen's have looked for a link between radon levels and lung cancer mortality, and the results have been inconclusive (no increased risk seen) when looking at the relatively low radon levels found in most homes.[9] However, some uranium miners have received doses that were 250 times higher than that of the average homeowner, and a positive correlation has been found between radon and lung cancer mortality for the higher radon levels experienced by these workers.[10] But we can't extrapolate those results to residential radon levels without knowing whether the LNT hypothesis is correct.

Epidemiological studies generally look at lung cancer mortality rates of *individuals* based on their exposure to radon, their age, their smoking habits, and other possibly relevant demographic variables. These studies typically use the case-control method in which the incidence of lung cancer mortality for people exposed to a given radon level is compared to that of a demographically matched control group. In contrast, Cohen's study is an example of an "ecological study," which looks at radon levels and mortality figures on a countywide averaged basis rather than for individuals. As a result, it gives much smaller statistical error bars than an epidemiological study because of the averaging over all persons in a county and the large number of data points (the 1,729 counties). But this strength of the ecological studies is also coupled with a significant weakness: by averaging over all individuals in a county, we may be ignoring other variables that are correlated with lung cancer, such as smoking status and age—a point we shall return to later.

Cohen's study (and ecological studies in general) has also been attacked by epidemiologists because of what has been called the ecological fallacy—the false belief that the average risk of a group of people can be found reliably from the average dose they receive. The average dose *may* determine the average risk in certain special cases, but not in general. For example, consider the following rather artificial example of two counties, A and B. In A all residents receive 9.99 cGy, and in B one resident receives 10.1 cGy and the rest receive 0 cGy. If there were a threshold of 10 cGy for lung cancer mortality, and radon were the only consideration, county A would have zero mortality due to radon, and county B, with one fatality, would have the higher average mortality. But, of course, the mortality rate of county A would be far higher than that of B if the LNT hypothesis

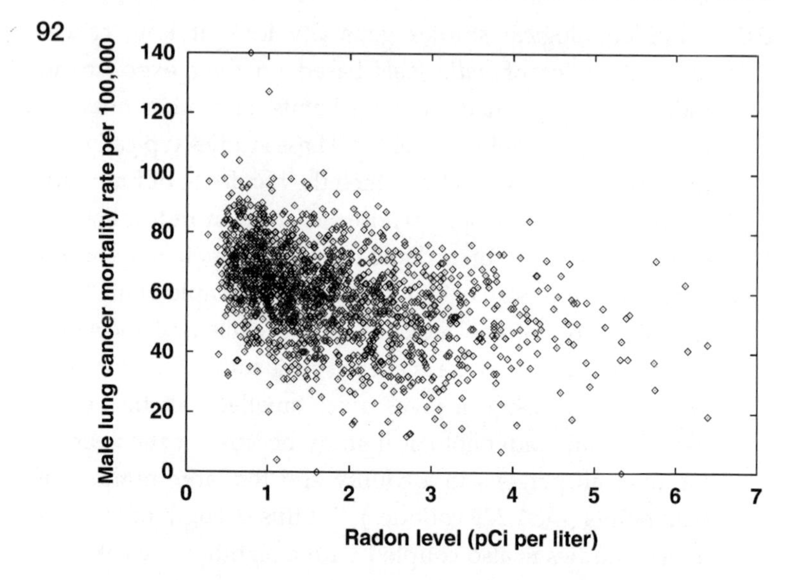

Figure 5.7. Lung cancer mortality for males per 100,000 persons versus mean indoor radon level in county of residence in picoCuries per liter of air, from reference in endnote [7].

were correct and no threshold existed. So, merely by looking at the mortality figures for each county, we cannot infer what is the correct dose-response relation for *individuals*.

Cohen has claimed that the ecological fallacy does not apply to his study, but others have challenged this claim. Let's first look at Cohen's results (see figs. 5.7 and 5.8) before getting into the claims and counterclaims. Figure 5.7 shows lung cancer mortality for males (per 100,000 people per year) versus the countywide average radon level r in units of picocuries per liter abbreviated pCiL^{-1}. This unit refers to the measured concentrations of radon gas in the air in county homes. (If you lived fifty years in a place where the radon level was one picocurie per liter, you would receive a dose of perhaps 5 to 10 cSv.)

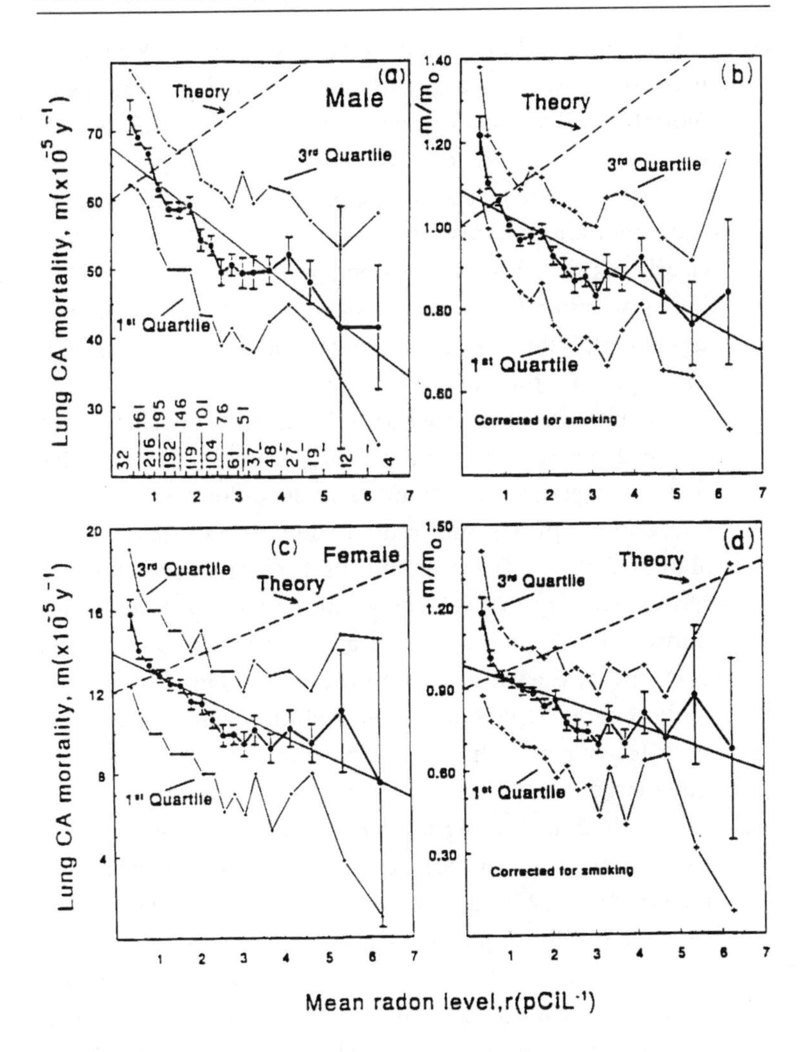

Figure 5.8. Lung cancer mortality per 100,000 persons versus mean indoor radon level in county of residence in picoCuries per liter of air from reference in endnote [7]. Graphs (a) and (c) are for males and females, respectively. Graphs (b) and (d) are the same as (a) and (c), except they have been corrected for smoking incidence variations. (Reproduced from *Health Physics* with permission from the Health Physics Society).

 Each dot in figure 5.7 represents one of the 1,729 counties used in the study. Although there is a widespread range in mortality (m) values for any given radon (r) level, we see that r and m appear to be correlated, with higher county-averaged radon levels corresponding on the average to *lower* mortality values. This correlation can be seen more clearly in figure 5.8, where Cohen has combined all counties falling within a narrow range of radon values. The quantities being plotted on the horizontal and vertical axes in figures 5.8a,c are the same as in figure 5.7.

The numbers of counties in each r-interval are shown just above the x-axis in figure 5.8. For example, 216 counties had the most common county-average radon level between 0.75 and 1.0 pCiL^{-1}, and only 4 counties had radon levels above 6.0 pCiL^{-1}. The size of the vertical error bars (showing the uncertainty in the average mortality rates) are determined both by the number of counties at any given radon level and by the variation in the mortality rates. Thus, at the highest radon levels in Cohen's study, where there were very few counties having such levels, the error bars are quite large.

In each of the four figures 5.8a–d, Cohen also shows a variety of curves: (1) a dark segmented curve connecting the data points; (2) curves labeled 1st and 3rd quartile (showing the first and third quartile of countywide mortality figures for any given radon level); (3) a downward sloping straight line, showing a best straight line fit to the data; and (4) an upward sloping dashed line labeled "Theory," which shows the prediction of the LNT hypothesis based on radon studies done at higher doses involving miners. (Studies of lung cancer in miners show that your risk of dying from lung cancer would double if you were exposed to around 9 picoCuries per liter over a lifetime.)

 It is interesting that the solid and dotted straight lines in figure 5.8 have about the same slope but opposite sign. Taken literally, that coincidence would seem to imply that, at low doses, radon is just about as beneficial as it is believed to be harmful at higher doses. Figures 5.8a and b are the data for males, and figures 5.8c and d are for females. Figures 5.8b and d are Cohen's attempt to correct 5.8a and c for variations in smoking prevalence from county to county. The smoking-corrected data look almost the same as the uncorrected data, which implies that Cohen believes that the smoking correction is small, and that it, therefore, cannot account for the negative correlation between radon levels and lung cancer mortality.[2]

It is vital that Cohen corrected his data properly for variations in smoking incidence if he wants to draw any valid conclusions from them. Smoking is by far the leading cause of lung cancer, because smokers have roughly twelve times the lung cancer mortality of nonsmokers. Therefore, a correlation between radon level and smoking incidence could possibly produce the kind of anomaly Cohen sees. This could happen, for example, if counties with lower radon levels just happened to have larger percentages of smokers. Moreover, it is possible that Cohen may not have properly corrected his data for smoking incidence, primarily because county-level data on smoking incidence were not available. Instead, Cohen had to use state-level data (based on cigarette sales taxes collected) and extrapolate such state-level data to the counties within a state by taking into account

[2] Note, however, the different vertical axes Cohen uses for the smoking-corrected data: the dimensionless quantity labeled m/m_0 is the ratio of the smoking-corrected mortality at any given radon level to its value at a radon level of zero.

 demographic variables such as the degree of urbanization. Such an extrapolation is problematic, given the large county-to-county incidences of smoking within each state.

Cohen's study has been challenged on precisely such grounds by a number of researchers, including Jay Lubin, a leading epidemiologist at the National Cancer Institute.[11] Let's look at Lubin's challenge in some detail and see whether it can explain Cohen's result. Lubin shows how it is possible for the relationship between dose and response to have opposite signs for individuals and groups. In other words, even if more radon produces more cancer in individuals, it can happen that counties with higher radon levels yield lower cancer mortality, even if the LNT dose-response relation applies.

The key to understanding Lubin's criticism is to realize that Cohen when he correlates mortality with county-averaged radon level, is using the wrong variable in his analysis. The county-averaged radon level that Cohen uses weights all persons in a county equally, but radon is much more damaging to smokers (whose lungs are already damaged) than to nonsmokers. Therefore, Lubin shows that when finding the correct average for a county, it is necessary to weight smokers much more heavily than nonsmokers because they are at much greater risk of dying from *radon*-induced lung cancer. In other words, rather than use a county-averaged measure of radon level, r, that weights all persons equally, Cohen should have used a value r' that weights the radon values found in smokers' homes twelve times more heavily than that found in nonsmokers' homes. Since smokers tend to live in homes with lower radon levels than nonsmokers (possibly because they air out their homes more), r and r' will, in general, have different values for a county.[12]

The real question, however, is, how large an error does Cohen make by using r instead of r', and is it large enough to cause his strange result? Lubin seems to believe that this error can lead to an error of an arbitrarily large amount.[11] But he comes to this conclusion through considering a highly artificial example of only two counties with very similar radon levels, for which he finds that r can be larger for the first county, yet r' can be larger for the second county. If we were attempting to find the slope of the dose-response curve using only two counties, such a reversal would reverse the sign of the slope. In a more realistic example, however, it is unclear whether the error Lubin has identified would give a large enough error to explain Cohen's result (i.e., that higher radon levels yield lower lung cancer mortality). Cohen believes that it would not (and I am inclined to agree with him).

For example, Cohen found that smokers' homes had radon levels, on the average, that were 90 percent those of nonsmokers' homes. If this figure applies to all counties equally, it can easily be shown with the aid of a little algebra that the ratio of the weighted and unweighted county-averaged radon levels (r'/r) will lie within 2 percent of its minimum value (0.944) for countywide smoking incidences in the range 10 to 60 percent.[12] If we were to apply this correction to Cohen's data, it would amount to shifting the x-coordinate (r) of each of the points in figure 5.7 to the left by an amount that varied between 0 and 2 percent of its value. From inspection of figure 5.7, it is clear that such a small correction would have no significant effect on the negative correlation between lung cancer mortality m and radon level r in Cohen's data.

What is the bottom line on Cohen's study? It is possible that the confounding effect of smoking and its correlation with radon level may explain Cohen's strange result, although specific attempts to do so seem to give too small a correction. (In defense of Cohen's study, it should be noted that he also analyzed his data on a statewide basis, where the smoking variations from state to state are known, and he still finds a negative correlation between *m* and *r*.) Critics of Cohen's study have found additional problems with it, including his assumptions regarding the fraction of one's life spent in the residence for which radon levels were measured, and the variations of smoking incidence with time.[13] Thus, we cannot be sure that some other correlation or combination of effects might not be able to explain Cohen's result.

By itself, this study cannot be said to demonstrate the reality of radiation hormesis. However, for the sake of argument, suppose we did assume that Cohen's result was due to hormesis. Would it then be in conflict with other studies of radon at residential levels showing no such effect? Jay Lubin claims that it would be in conflict with them, as shown by figure 5.9.[11] But this claim is mistaken. Cohen's data give error bars so large as to make his results meaningless for radon levels above about 4 $pCil^{-1}$ (or 150 Bq/m^3). Therefore, it is quite misleading to extend the curve labeled "Cohen's estimate" beyond 150, as Lubin has done. If Cohen's result were due to hormesis, it is *not* in conflict with other residential studies, and certainly not with miner studies.

Theoretical and Policy Considerations

Although we have seen that existing data do not convincingly show that the LNT hypothesis is wrong, or that either hormesis or a threshold occurs, a number of theoretical arguments have been advanced for these latter two possibilities and against LNT.[14] Many of these ideas involve biological defense mechanisms, whose efficacy can be enhanced by low doses of radiation and which prevent cancers from being developed even after they have been initiated by a radiation dose. For example, data exist which apparently show that a low dose of radiation administered before a much higher dose appears to decrease the extent of genetic damage done by that higher dose.[14] The mitigation of harm done when radiation doses are spread out in time also implies that biological defense mechanisms are important. A discussion of these ideas and references to relevant studies can be found in Cohen's paper.[15] We do not

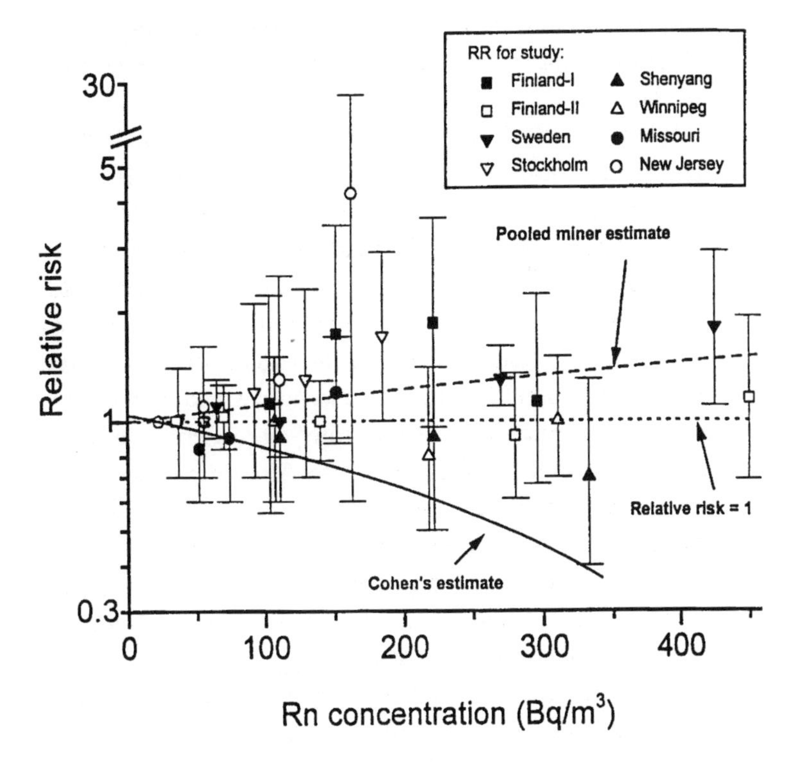

Figure 5.9. Relative risk of lung cancer mortality versus indoor radon level in air from eight radon studies according to the reference in endnote [11]. (Reproduced from *Health Physics* with permission from the Health Physics Society.)

 dwell on them further here, because mechanisms explaining the basis for a threshold or hormesis would seem to be of limited relevance if the phenomena have not been clearly shown to exist.

It would seem worthwhile to try to resolve the nature of the dose-response function at low radiation levels in view of very important policy implications. For example, very considerable sums of money would need to be spent to clean up the radioactive wastes from nuclear weapons plants and the storage of wastes from nuclear power plants. However, if low levels of radiation are harmless (or conceivably even beneficial), most of that money would be wasted, since the plans for the waste disposal would not need to be as elaborate.

Likewise, the Environmental Protection Agency (EPA), which has estimated that between 5,000 and 20,000 lung cancer deaths are due to radon (out of a total of 130,000 annually), currently warns homeowners to take remedial action against radon at levels above 4 $pCil^{-1}$. But at such radon levels, the harm is entirely theoretical. The EPA estimate assumes LNT is true and Cohen is wrong. (If Cohen were right, radon should be enhanced in homes, rather than eliminated.) It is unclear what the "safe" or conservative course of action would be in such a case. Most homeowners are probably more trustful of the EPA than some scientist proposing a crazy-sounding idea that radiation might actually be good for you at low doses. Nevertheless, it seems regrettable that the EPA may have engaged in scare tactics in the past to promote greater awareness of the radon problem, even though the existence of the problem has not been scientifically established, only merely extrapolated from higher doses, assuming LNT is valid.[15]

Similarly, many people in the nuclear industry believe that the public has long had a radiation phobia partially fed by the LNT hypothesis and the corresponding idea that there is no "safe" level of radiation. One indication of that radiation phobia is the public concern over food irradiation (for the purpose of killing bacteria), which results in essentially zero radioactivity in the food after irradiation. In the past, public concern over radiation and nuclear power has led to greater safety margins and new inherently safe reactor designs, which is all to the good. Carried to their extreme, however, excessive nuclear fears may limit society's future energy options and result in higher energy costs.

My rating for the idea that low doses of nuclear radiation are beneficial is 1 cuckoo.

6 The Solar System Has Two Suns

ON THE next clear moonless night, here's a vision test you might want to try. One of the stars in the Big Dipper is actually a double star or a binary—two stars in close orbit about each other. See if your eyes are keen enough to tell which one is the double star. Roman generals at the time of Julius Caesar apparently used this vision test for their soldiers. Surprisingly, even though most stars appear to the naked eye as single points, binary stars are actually quite common. In fact, astronomers suspect that perhaps even a majority of stars are binaries, although they don't know for sure, since many companion stars may be too small to detect.

What would it be like to live on a planet in a solar system with two suns? Two possibilities come to mind. The first would be that the two suns orbited each other in a tight orbit, and our planet revolved in a larger orbit about their center of mass. As our planet rotated on its axis, the two suns would appear to move in tandem during their daily journey across the sky. But, unfortunately, such planetary orbits about a double sun are not stable, so we wouldn't be able to enjoy the double sunrises and sunsets for very long.[1] A second more promising possibility would be that our planet orbited one of the suns, while the two suns orbited each other in an orbit of much greater size.

[1] Actually, the planet's orbit could be stable if its distance to the double sun were very much greater than the distance between the suns; but in that case, it probably wouldn't receive enough sunshine for life to be able to exist on its surface.

For this type of orbit, one sun would be much more distant from our planet than the other, and it might not greatly disturb our planet's orbit. Depending on the second sun's brightness, it might be indistinguishable from the background stars when viewed from our planet. Believe it or not, some scientists think that we actually live in such a double-star system, though no one has yet spotted Sol's companion among the stars. Of course, that fact does not rule out the possibility of a solar companion, because the star could be too dim to be seen, as is the case with many small "brown dwarf" or "black dwarf" stars.

What Killed the Dinosaurs?

The evidence for a possible companion star to the sun comes from a study not of the heavens, but of the Earth's rocks, from which geologists have learned to read the history of the Earth. The science of geology was founded on the belief that the structures found on the present-day Earth should be explainable on the basis of processes and phenomena observed to occur on Earth today. The present day geological record should be understandable when these processes, such as erosion and weathering, slowly operate over many millions of years. This belief, known as "uniformitarianism," was in reaction to the idea of sudden change brought about by catastrophes, which geologists of an earlier era rejected as being outside the bounds of their science—a carryover from an earlier prescientific era or from a literal reading of the Bible.

The geological record of much of the Earth's history is based on the sequence of layers of sediments deposited to form rocks, and on the fossils each layer contains. These various rock layers follow a common sequence in most places of the world, and hence they can be used to order successive geological periods. One interesting geological boundary that occurred 65 million years (abbreviated Myr) ago is between the end of the Cretaceous period and the beginning of the Tertiary period. The Cretaceous-Tertiary boundary, usually called the K-T, marks the end of the age of the dinosaurs. (Logically, you might expect the boundary to be called C-T, not K-T, but the letter C is already taken for the Cambrian period.)

At one time, many theories existed as to why the dinosaurs died out, including the notion that they were unable to adapt to a changing climate, which is why we now use the pejorative term "dinosaur" for people who are unable to adapt. The real story of why the dinosaurs became extinct, however, puts these animals in a more favorable light—which seems only fair, since they lived on Earth eighty times as long as humans have so far existed. The mystery was uncovered by a young geologist, Walter Alvarez, who dared to ask a question that challenged the very basis of his discipline. Alvarez wanted to know whether the extinction time was gradual or abrupt—and therefore due to a catastrophe—an idea that most geologists ridiculed. (Bear in mind that terms like "gradual" and "abrupt" are used very differently in geology than in human affairs. In geology, a change that takes place in a time significantly less than a million years would be considered abrupt.)

Alvarez's question about the dinosaur extinction was also a question about many other life forms that existed at the end of the Cretaceous period. The species of tiny fossils below the K-T boundary layer are completely different from those above it; thus the boundary appears to mark the extinction of not just the dinosaurs, but also about 70 percent of all species then existing. It was what geologists now refer to as an example of a mass extinction. Alvarez found that a one centimeter thick layer of clay appeared right at the K-T boundary, which was free of fossils. Enlisting the aid of his father, the renowned physicist Luis Alvarez, he hoped to solve the mystery of the dinosaur disappearance by analyzing this thin clay layer. Essentially, the father-son team wanted to find if the clay layer was deposited slowly over a very long time or relatively quickly.

In order to answer their question about the K-T boundary clay, they analyzed it for the presence of the element iridium, a very rare element on the surface of the Earth. Why

"All I'm saying is now is the time to develop the technology to deflect an asteroid."

Figure 6.1. © The New Yorker Collection 1998. Frank Cotham from Cartoonbank.com. All Rights Reserved.

 iridium? That particular element is continually being slowly added to the Earth's surface by infalling dust from space. If the clay layer had been deposited over a long time, it should contain a lot of iridium, whereas if it was deposited in a short time, it should contain very little.

What the Alvarez team found was very surprising. The clay in the K-T boundary layer contained much more iridium than could be explained even under the slow-deposition scenario. Also, the high iridium content was present only in the clay layer, but not above or below it. The initial iridium report was followed up by other findings of a high iridium content in the thin K-T boundary layer all over the world, showing that the high iridium level was not a local anomaly. What did it mean? After much thought, the Alvarez team realized that the iridium signal was strong evidence for a large meteorite impact.[1]

Iridium is 10,000 times more common in meteorites than on the surface of the Earth. Even though the original Earth had the same abundance of iridium as meteorites, little of that original iridium would be found on the Earth's surface. That's because of iridium's affinity for iron, which caused most of the iridium on the early molten Earth to sink, along with iron, toward the center of the Earth. Based on the amount of iridium present in the K-T boundary clay, the Alvarez team was able to estimate how much iridium was deposited worldwide. Then, knowing the abundance of iridium in meteorites, they could deduce the approximate size of the meteorite. Their estimated diameter was 10 kilometers, about the height of Mount Everest. Large meteorites can wreak havoc on the planet, not simply because of their large size, but more importantly, because of their extremely high speed (perhaps 50 kilometers per second or 100,000 miles per hour), which results in a tremendous explosion when much of the meteorite vaporizes on impact.

But why was the iridium layer deposited worldwide? On impact, vaporized parts of the meteorite, along with ejected material from the impact site itself, would rise high above the atmosphere to later fall back to Earth all around the globe. If the meteorite was large enough, the impact would be sufficiently catastrophic to destroy most life on Earth. The exact killing mechanism is uncertain, but there are many likely possibilities. Any life within a few hundred kilometers of the impact site would have been killed directly from the blast created when the meteorite vaporized (estimated to be the equivalent of a million times the total world's nuclear arsenals). Additionally, if the impact occurred in the ocean, it would create a tsunami (tidal wave) that would be over a kilometer in height when it hit land.

The leading candidate for a cause of death on a worldwide basis is the extensive dust and smoke that would be injected into the upper atmosphere, where it could remain for several years before settling back to Earth. Assuming a meteorite 10 kilometers in diameter, the worldwide pall of dust (and smoke from worldwide wildfires) would block sunlight enough to prevent photosynthesis and cut the food chain off at its base, leading to worldwide starvation. (A similar "nuclear winter" scenario has been suggested as the possible outcome of a nuclear war, although it is unclear whether the nuclear version would be nearly as catastrophic as first suggested[2].)

When the Alvarez team put forth their hypothesis for a meteorite impact as the cause of the K-T mass extinction, it was greeted by considerable skepticism. Although there was much evidence for meteorite cratering on many bodies in the solar system, including the moon, many scientists thought that impacts large enough to catastrophically affect the Earth's environment only occurred when the solar system was very young. They found it difficult to accept the idea that Earth experienced a catastrophic meteorite impact as recently as 65 Myr ago. In large part, this skepticism was based on the ingrained belief of geologists in the uniformitarianism idea, which allowed no room for catastrophic events.[2][3]

There were also many objections to the impact hypothesis on legitimate scientific grounds. For example, it was noted that an episode of vulcanism might be the cause of the mass extinction since volcanoes also release iridium. However, this alternative hypothesis might explain the iridium, but it didn't explain other rare elements in the platinum group that were later found in the K-T clay. These other elements were found in relative abundances that matched those found in meteorites, but not in volcanoes. The K-T clay also contained clear signs of the extremely high temperatures and pressures present during a meteorite impact (or a nuclear explosion), but not during volcanic eruptions. These signs include the presence of shocked quartz as well as microtektites, the tiny droplets of melted glass formed during impact.

Perhaps the greatest shortcoming of the impact hypothesis was the lack of evidence for any known crater of the requisite size and age to match the K-T mass extinction. The Alvarez team realized that they would need to locate the crater before their hypothesis could gain acceptance. After many false leads and a ten year search, the Chicxulub crater was finally located in 1989 adjacent to the Yucatan peninsula of Mexico. The date of the underwater crater was found by radioactive dating methods, and it matched that of the

[2] Eugene Shoemaker has pointed out that there is so much evidence for the occurrence of large meteorite impacts during the last 500 Myr that impacts can now easily be absorbed into the uniformitarian doctrine.

K-T boundary (65 Myr). Also, its 200 kilometer diameter was consistent with what would be expected, given the explosive force created by the impact of a 10-kilometer-diameter meteorite. Moreover, the crater bore all the earmarks of one caused by an impact rather than vulcanism (including a raised ring in the middle, which is also seen in nuclear bomb craters).

Finally, as expected, the thickness of the K-T layer was greater the closer you got to the crater site. Although there were still holdouts, for many researchers the Chicxulub crater was the "smoking gun" that showed that the K-T mass extinction was caused by a meteorite impact. (Oddly enough, the Chicxulub site had been identified as an impact crater by geologists working for the Mexican national oil company PEMEX many years earlier, but this was not widely reported.) One further piece of evidence that large meteorite impacts were not restricted to the early years of our solar system was supplied in 1994, when the Shoemaker-Levy comet slammed into the planet Jupiter. The force of that impact, had it occurred on Earth, would have been enough to wipe out most life on the planet.

Once it had been shown that the K-T mass extinction had been associated with a meteorite impact, scientists wondered about other mass extinctions—could they all have been caused by impacts? The geological record of the last 540 Myr contains evidence for five major mass extinctions, and perhaps twenty minor ones. In the largest mass extinction that occurred 250 Myr ago, perhaps 96 percent of all species were wiped out. None of the other mass extinctions have been investigated as extensively as the K-T, but a number of them do show the same telltale signs of meteorite impact, including the presence of craters containing shocked quartz and microtektites. Additionally, craters were found that match closely the ages of three of the big-five mass extinctions, plus a couple of the smaller ones that occurred during the last 250 Myr.[4] There would seem to be compelling evidence that the major mass extinctions (if not all of them) are associated with impacts—although the possibility remains open that other causes, including vulcanism, may have been contributing factors.

Are Mass Extinctions Periodic?

Prior to the discovery that the K-T mass extinction was caused by a meteorite impact, a young geologist named John Sepkoski was looking at the fossil evidence to see how the rate of extinctions of species varied with time. The extinction rate of species can be defined as the number of species becoming extinct in a given time period expressed as a fraction of the number then existing. Biologically, species are grouped into genera (plural for genus), and genera are grouped into families. (For example, we are members of the species *Homo sapiens* that belongs to the genus *Homo* in the primate family.) To reduce the random fluctuations or "noise" in the extinction data, Sepkoski looked at the extinction rate of families of organisms rather than species. (100 percent of every species in every genera would have to die out before a family would be considered extinct.) Figure 6.2, taken from a 1984 article by David Raup and John Sepkoski, shows how the family extinction rate has varied with time over the last 250 Myr.[5]

The extinction peaks in figure 6.2 are of greatly varying heights, and one might question whether some of the peaks are genuine or simply noise. Nevertheless, one striking feature of this plot does seem to stand out: the peaks (with a few exceptions) seem to be equally spaced in time.

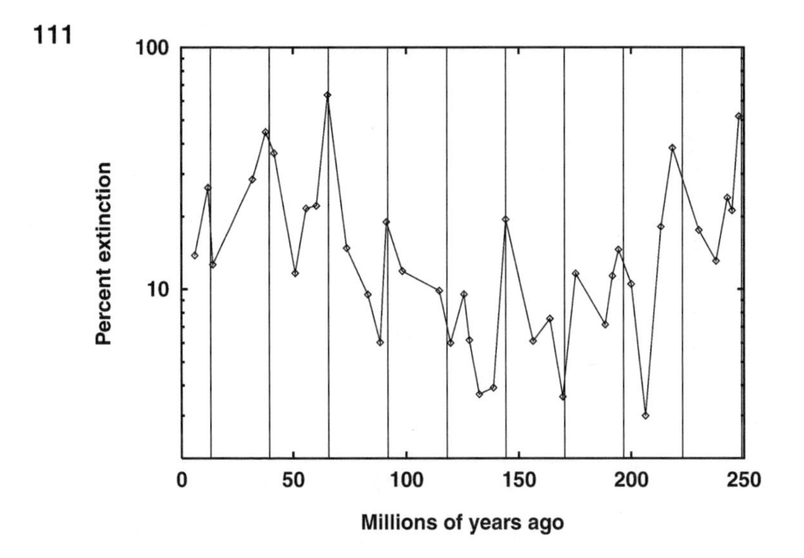

Figure 6.2. Extinction record for the past 250 million years, according to reference in endnote [5]. The vertical lines correspond to a best-fit 26.2 million year cycle. The extinction rate is the number of families becoming extinct in a given geologic time period expressed as a percentage of the number then existing. Note the use of a logarithmic vertical scale.

The vertical lines drawn at 26 Myr intervals would seem to fall quite close to at least eight of the peaks in figure 6.2. Raup and Sepkoski used a variety of statistical techniques to see whether this degree of agreement might have occurred by chance. They estimate that random data would show a 26 Myr periodicity of the kind seen in figure 6.2 with a probability $p < 0.0001$, and that random data would show *any* period with $p < 0.01$. (The latter probability estimate is the more reliable one when judging whether the result could be due to chance, because we have no expectation of any particular period.) A level of confidence of $p < 0.01$, while not overwhelming, offers some support that the regularity seen in the figure is statistically significant.

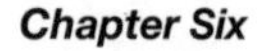

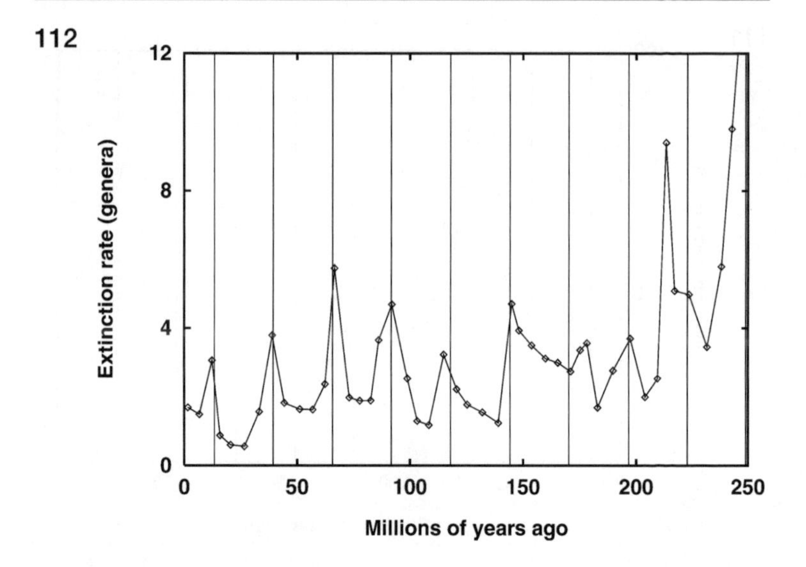

Figure 6.3. Genus level extinction rate during the last 250 million years, according to reference in endnote [6]. The vertical lines correspond to a best-fit 26.2 million year cycle.

In a subsequent 1989 publication, Sepkoski reanalyzed the data (see fig. 6.3) and reported a result with even greater statistical significance.[6] In his 1989 article Sepkoski used extinction at the genus level rather than the family level. Note that many of the extinction peaks now appear much sharper than before. For these revised data, Sepkoski cites a probability of a chance agreement with a 26.2 Myr period to be only around 10^{-6}. One way to understand how he obtains such a small probability is to look at how close each peak comes to an exact equal spacing in time.

This can be easily done using a circular graph (see fig. 6.4), where 12 o'clock on the circle corresponds to the present, and each 26.2 Myr backward in time represents a full 360 degrees clockwise rotation. Thus, for example, a peak which occurred either 13.1 or 39.3 Myr ago (or any multiple

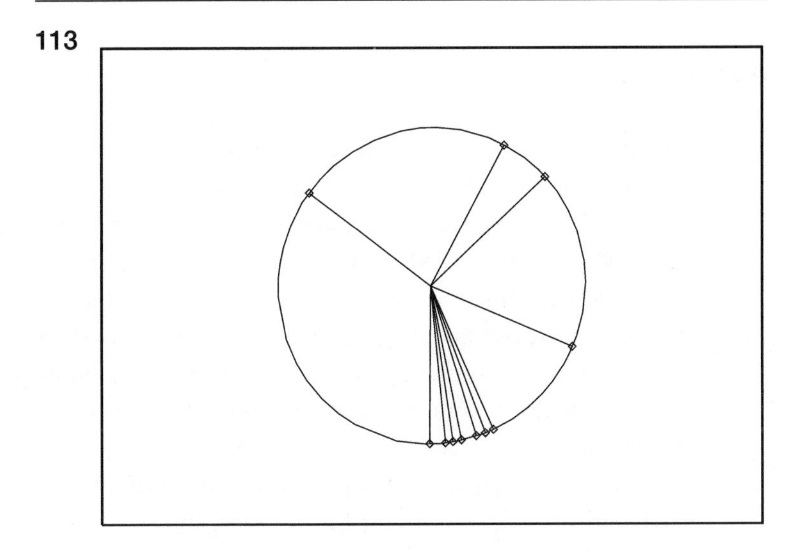

Figure 6.4. A circular graph of the ages of eleven generic extinction peaks in figure 6.3. The present is 12 o'clock on the circle; each 26.2 million years backward in time represents one complete revolution. Note the clustering of seven peaks in an angular region near 6 o'clock.

of 26.2 plus 13.1 Myr) would be shown at six o'clock on the circle. The noteworthy aspect of figure 6.4 is the strong clustering of peaks in a narrow angular range (24 degrees wide), which is indicative of a real 26.2 Myr periodicity.

How unlikely would it be to find seven out of eleven randomly chosen angles falling within a 24 degree arc? The chances of one randomly chosen angle lying within a particular 24 degree arc is 1/15, since 24 degrees is 1/15 of a full circle. Based on the laws of probability, the chances of seven out of eleven randomly chosen angles falling inside the 24 degree arc is given by $p = (1/15)^7\ (14/15)^4\ 11!/(7!4!) = 1.5 \times 10^{-6}$, which is close to the probability estimate quoted by Sepkoski. ("!" represents the factorial function, for example, $4! = 4 \times 3 \times 2 \times 1$.)

The Nemesis Hypothesis

If we couple a 26.2 Myr periodicity in mass extinctions with meteorite impacts as their cause, we must conclude that showers of large meteorites regularly bombard the Earth at 26.2 Myr intervals. What could be the cause of these periodic "death showers"? One type of meteorite is comets, which are believed to originate from the Oort cloud, a spherical cloud surrounding the solar system at its outer edges. Suppose the sun had a companion star that completed its orbit every 26.2 Myr, and suppose that the orbit was sufficiently eccentric (elongated) to pass through the Oort cloud once each orbit.[7] In that case, the star might dislodge a number of comets from the cloud and send them into orbits heading toward the inner solar system and into possible collision trajectories with Earth at 26.2 Myr intervals.

A name has been suggested for the hypothetical star responsible for this killing machine: Nemesis, the mythic Greek goddess of vengeance who punishes the excessively rich, proud, and powerful. The paleontologist Stephen Jay Gould has argued that a more appropriate name for the solar companion star would be Shiva, the dancing Hindu god of destruction, who punishes his victims at random, whether or not they deserve it. Perhaps the issue of the star's name might best wait until such a star is actually found.

Some scientists have suggested that the star will never be found because the reason for the periodicity in meteorite impacts lies elsewhere. One alternative hypothesis involves the motion of the solar system through the Milky Way galaxy—a giant, disk-shaped collection of 100 billion stars.[8] In addition to its orbit around the center of the galaxy, the sun also oscillates up and down through the central plane of the galactic disk. (Just picture the sun as one of those carousel horses that bob up and down as the carousel-galaxy rotates, and you'll get the idea.) These oscillations are estimated to have a period of about 66 Myr, meaning that the solar system passes through the galactic plane each half period of about 33 Myr.

It is possible that the extra concentration of matter in the galactic plane might cause comets to be dislodged from the Oort cloud at 33 Myr intervals when the solar system passed through the galactic plane. A 33-Myr interval is sufficiently close to 26.2 to give this explanation some degree of plausibility. However, there are also some problems with this alternative explanation for the periodic death showers. One problem is that the solar system is currently very close to crossing the galactic plane, and yet we are currently roughly halfway between mass extinctions rather than in the midst of one. In addition, Sepkoski's mass extinction peaks are extremely sharp, and yet we would expect a gradual rather than a sudden fall-off in concentration of material as we moved away from the galactic plane. (Oscillations through the disk should be more like passing through a cloud than like going back and forth over a speed bump.)

The original explanation for the cometary death showers (Nemesis) also has some problems with it. One difficulty is that the star would have to have some unusual properties. Given a period of revolution of 26.2 Myr, the semimajor axis of the star's elliptical orbit would have to be 1.4 light years, and fewer than 0.1 percent of binary stars are observed to have orbits that large.[7] In addition, Nemesis could be neither too massive nor too light. If it were more massive than about one percent of the sun's mass, it would cause enough of a disturbance in the orbits of the planets to be noticed.[9] Yet if its mass were much less than one percent of the sun's mass, it wouldn't cause enough of a disturbance in the Oort cloud to dislodge a sufficient number of comets.[9] Taken together, these three restrictions on the properties of Nemesis cast some doubt on whether it really exists. An analogy may help clarify the last statement.

Suppose a jealous spouse hires a detective to find out if his wife is cheating. The detective stakes out the couple's home while the husband is out of town on a number of trips, and he also taps the couple's phone and follows the wife's car while she drives around town. Following the surveillance, the detective reports that the wife (a) received no suspicious phone calls, (b) had no male visitors, and (c) never drove to any motels. Do these observations exonerate the wife? Of course not. She could still be cheating and might have (a) met her lover somewhere other than at a motel, (b) dated a lover who doesn't use a phone, (c) dated a female lover, (d) spotted the detective and stayed away from her lover, as well as any number of other possibilities. Even though the detective's negative report doesn't prove the wife wasn't cheating (no report could possibly prove a negative), it certainly casts some doubt on the possibility. Using much the same logic, we can say that if the only type of Nemesis that would have escaped our notice is one with an unusual mass and orbit size, we have grounds to question the star's existence. But there is an even more important reason for doubting that Nemesis really exists. The only reason for suggesting Nemesis in the first place was to explain the periodicity in meteorite impacts. If the evidence for that periodicity is lacking, the Nemesis hypothesis loses all its rationale. In Raup and Sepkoski's original analysis of extinction peaks, they closely looked at data only through the last 250 Myr. More recently, Sepkoski analyzed extinction data going back more than twice that far. The clean 26.2 Myr spacing of the extinction peaks does not appear in the

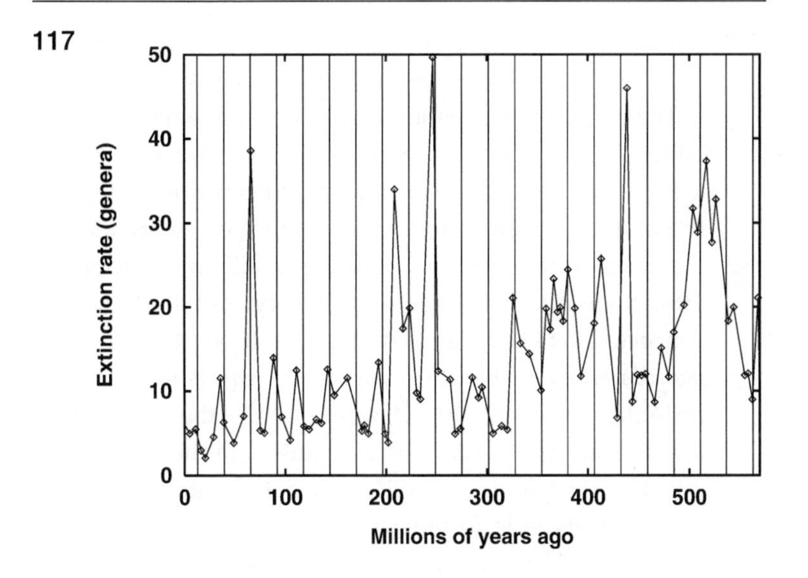

Figure 6.5. Percentage of extinction by genera during the last 570 million years, according to reference in endnote [10].

data any farther back than 250 Myr (see the equally spaced vertical lines in fig. 6.5).

Both the fossil record and the geological timescale become less reliable before 250 Myr ago, but a gradual degradation in the data beyond this time probably couldn't explain the relatively abrupt failure of the regularity. Some authors (Rampino and Haggerty) have tried to argue that the peaks in figure 6.5 are really periodic beyond 250 Myr, but with a slightly longer period.[10] However, their analysis relies on making arbitrary selections in the data. For example, they report a clear 27.3 Myr periodicity if they analyze the data only for the last 515 Myr, but not for the entire 540 Myr period. Moreover, even with this arbitrary selection, the probability they report of finding a periodicity due to chance ($p = 0.02$) is considerably greater than Sepkoski reported when analyzing the data only up to 250 Myr. Real regularities almost always become more—not less—statistically significant when the amount of data is increased.

The breakdown of periodicity when more data have been added is reminiscent of many other regularities that went away when new data were added. One famous example is the well-known Titus-Bode rule for the average distances of the planets from the sun. Like Sepkoski's periodicity, the Titus-Bode rule follows from no theory, and it was simply concocted to fit the observed distances. It says that the average distances of the nth planet to the sun is given by the formula: $r_n = (a_n + 4)/10$ AU, where 1 AU is the Earth's average distance to the sun. The constants a_n in the formula are given by $a_1 = 0$ and $a_n = 3 \times 2^{n-2}$ for $n > 1$. For the three innermost planets, this formula predicts $r_1 = 0.4$ for Mercury, $r_2 = 0.7$ for Venus, and $r_3 = 1.00$ for Earth.

When the Titus-Bode rule is applied to each of the planets known at the time the rule was formulated (all but Neptune and Pluto), it gave excellent agreement: the average discrepancy between the predicted and observed orbit sizes were only 2.8 percent. Unfortunately, the rule failed badly after Neptune and Pluto were discovered, giving discrepancies of 23.5 and 48.9 percent for these last two planets. (Oddly enough, however, if Neptune didn't exist, Pluto would be almost at the right distance for planet number eight.)

Nowadays, most astronomers regard the Titus-Bode rule as an interesting oddity, whose excellent agreement with the first seven planetary distances is just a matter of chance. A minority of astronomers might wish to find a way to salvage the rule, if only there were some way to explain away Neptune's inconvenient existence. The situation here is similar to Sepkoski's periodic mass extinction hypothesis, which seemed to fit the data so well for the most recent 250 Myr, only to fail badly when new data beyond 250 Myr were added. One could imagine scenarios where a real periodicity was confined only to the 250 Myr period where it was first observed (for example, if Nemesis had been a passing star that was captured into a solar orbit 250 Myr ago), but a simpler hypothesis would be that the initially reported periodicity was just due to a chance agreement. It is clear that the human mind has a wonderful ability to see regularities in numerical data, even when none really exist.

Some readers may take exception to the last statement as it applies to the hypothesis of periodic mass extinctions. Given an infinity of possible mathematical formulas, Titus and Bode may have had little trouble concocting one that fit the planetary distances reasonably well, but surely it is quite another matter to find something as simple as regular spacing in time occurring on a chance basis. How could nature have created such a regularity by chance, if the probability of such a pattern occurring at random were only about 10^{-6}, as Sepkoski reports?[6]

One answer to the puzzle is that the probability against chance is actually not nearly as low as Sepkoski has reported, according to some authors who have reanalyzed the data.[11] Another reason for skepticism is based on Sepkoski's own analysis. In his 1984 paper with Raup, the probability of a chance 26 Myr periodicity was quoted as 10^{-4}, and for any period the probability was quoted as 10^{-2}. It was only in his later (1989) paper, after reexamining and refining the data (but not adding to it), that Sepkoski reported a much more significant result with the lower probability of around 10^{-6}. What was the nature of this refinement process that caused the extinction peaks to fit the periodicity model much better in the later analysis?

The peaks in figure 6.3 are much narrower than those in figure 6.2 because of the switch from extinctions at the family to the genus level. But the more significant change from figure 6.2 is in the positions of the peaks in time (based on a different choice of geological stages) and, most crucially, in the number of peaks that agree with a 26.2 Myr period. A close comparison of the peak positions in figures 6.2 and 6.3 reveals that 6.3 has one more peak (near 120 Myr), which lies very near a vertical line. This one additional peak would make a chance agreement with a 26.2 Myr periodicity about twenty times more unlikely than in Sepkoski's previous analysis.[3]

If some geologists should believe or suspect that the peaks were truly periodic, it is easy to understand how this belief could subtly influence their choices in evaluating extinction data to achieve better agreement with the periodicity. It is important to stress that this process of refining data to achieve better agreement with a "known" result need not reflect any sort of cheating, just the natural tendency of all analysts of data to tease out of the data what they expect to see.

Some researchers have suggested that adjustments in the geologic timescale or in the extinction data might easily make a periodic series of peaks more random, but it shouldn't make a random arrangement of peaks more periodic. But this suggestion only applies to adjustments that are truly random and is not informed by what one suspects to be the "correct" answer. In the latter case, it is quite plausible that adjustments in the data can make a random arrangement of peaks more periodic.

In conclusion, it would seem that the evidence for periodicity in mass extinctions (the sole basis for the Nemesis hypothesis) is open to serious question. This assessment could change tomorrow if astronomers should happen to spot

[3] The chance of six randomly chosen angles falling within a particular 24 degree arc is $p = (1/15)^6(14/15)^5 11!/(6!5!) = 3.0 \times 10^{-5}$, which is twenty times less than that of finding seven in that interval.

 Nemesis among the stars. If they only knew where to look, spotting it might not be too difficult; based on Nemesis's predicted orbit size, it should be closer than any other known star. Actually, in one automated search, distances were measured for all faint stars in the Northern Hemisphere.[12] While the search has turned up negative, it still leaves open the possibility that Nemesis is too dim to be seen with the system used, or that it isn't visible from the Northern Hemisphere.[12]

My rating for the idea that Nemesis exists is 2 cuckoos.

7 Oil, Coal, and Gas Have Abiogenic Origins

IF YOU run out of gas on the highway, it can be a minor annoyance; but if the world runs out of gas, the consequences could be catastrophic. How much oil and gas does the world have left? In the mid-1970s, some petroleum experts predicted that the world's oil reserves would run out in fifteen years, and the "energy crisis" was in part a consequence of those pessimistic predictions. Now, some twenty-five years later, the oil shortage seems not nearly as dire. Yet, many observers still warn that we are rapidly depleting the world's supply of hydrocarbon fossil fuels—especially petroleum, which is predicted to run out in a matter of decades not centuries.[1] Might the predictions be right this time?

Natural gas, petroleum, and coal are called fossil fuels because of their presumed biogenic origin in buried decaying vegetation that was deposited in layers and transformed through heat and pressure over the course of time into their present form. Geologists once vigorously debated the origin of hydrocarbons, but that debate was considered settled when these substances were found to contain traces indicating a biological origin. Sometimes, however, it is useful to reexamine long-settled debates when new information comes to light. It often takes an outsider to reopen the mat-

[1] According to the statistics division of BP Amoco, the world's estimated "proved reserve" would last forty years at current rates of consumption, or twenty-five years (until the year 2025), given the present annual 2 percent increase in oil consumption.

ter, since practitioners in a field may have too much at stake in the status quo to notice the gradual buildup of inconsistencies.

Thomas Gold, a well-respected scientist in his own right, is just such an outsider. Gold, a retired professor of astronomy from Cornell University, has a track record of coming up with weird ideas, some of which were shown to be correct after being initially rejected by the experts in a field.[2] One of Gold's weird ideas (attacked or ignored by most geologists) is that hydrocarbons are not the fossil fuels we believe them to be, but were part of the original composition of the planet, and that they are present in the deep crust and mantle of the Earth in far greater abundance than geologists believe. Gold has expounded on these ideas in several books and articles on which much of the material in this chapter is based.[1, 2, 3, 4] If Gold is right, the practical stakes are enormous, since it would mean that fears of an oil or gas shortage could be put off to the distant future. It would also mean that deep sources of natural gas and oil could be found at far more locations around the globe than those found to date.

What is the nature of the evidence that convinced most geologists of the biogenic origin of petroleum? First, petroleum is almost always found to contain certain groups of molecules that are only produced in the breakdown of living matter. Second, petroleum frequently exhibits "optical activity," meaning that polarized light passing through it has its plane of polarization rotated. This observation can be understood in terms of the kinds of molecules found in petroleum, which often come in mirror-image, right- and left-handed varieties, just like right- and left-handed screws.

[2] Gold has been shown to be right in such diverse areas as a theory of hearing, the nature of pulsars, and a theory of the Earth's axis of rotation.

 The phenomenon of optical activity shows that petroleum contains unequal numbers of right- or left-handed molecules. Here again we have an indicator of the effects of life, since living organisms have evolved to eat substances such as right-handed sugar (dextrose) but not its left-handed mirror image (levitose). A third indicator of the biogenic origin of petroleum is the predominance of molecules having an odd number of carbon atoms—another sign of processes involving living systems.

Why Reopen the Debate?

Gold points out that each of the preceding pieces of evidence for a biogenic origin of petroleum has a plausible alternative explanation in the light of new discoveries. Finding biological traces in petroleum need not point to a biogenic origin, but could equally well be explained based on a biological contamination of a hydrocarbon fluid coming up from great depth. At one time it was believed that the biosphere—the domain of living systems—ended a short distance beneath the surface of the planet. But it now appears that much of the Earth's crust may literally be teeming with life that thrives on extreme conditions of temperature and pressure. These "extremophile" bacteria have been found to survive at temperatures as high as 169°C in a marine sediment drilling core. At such high temperatures, bacteria could survive at depths up to around 10 kilometers, where the pressure is sufficient to permit water to remain liquid.[3] Gold has estimated that the total mass of

[3] In nonvolcanic regions subsurface temperatures rise about 15 °C for each kilometer in depth.

125 subsurface bacterial life may exceed the mass of all life on the surface.[3]

If Gold is correct about petroleum and other hydrocarbons seeping upward from deep in the Earth's crust and mantle, it is easy to understand how they would acquire biological traces as they passed upward through the Earth's crust, and that bacteria used them as a source of energy. In fact, one could even reverse the argument and say that if hydrocarbons did not continually come from deep inside the Earth's crust, underground bacteria would lack a plausible energy source for their long-term existence.[4]

Another new piece of information that warrants reopening the question of the origin of hydrocarbons involves theories about the formation of the Earth. At one time it was believed that the Earth was molten shortly after its formation. Given an initially molten Earth, any hydrocarbons present when the Earth was formed would be destroyed before the Earth solidified. Therefore, an abiogenic primordial origin of hydrocarbons seemed highly implausible given an initially molten Earth.

Current theories about the Earth's formation involve collisions of cold chunks of material yielding an Earth that was not molten throughout. Much of the Earth's mantle is quite hot, but not molten, and this fact has also been offered as a reason that hydrocarbons could not originate from great depths. But these objections fail to consider the stabilizing effects of high pressure, which would prevent oxidation of hydrocarbons if they originated from depths of up to 300 kilometers.[5]

[4] As spelled out in endnote [1], Gold believes that life originated inside the Earth rather than on the surface, where conditions initially would have been much less hospitable to life.

 A third reason for reopening the debate about the origin of hydrocarbons is that large amounts have been found throughout the solar system on every planet but Venus, Mars, and Mercury. They are also found on many planetary moons. Methane has been found most frequently, but ethane, other hydrocarbon gases, and tar have also been observed.[5] The absence of hydrocarbons on Mars and Mercury is due to the lack of a sufficiently dense protective atmosphere, and no information exists about surface hydrocarbons on Venus because of its dense, opaque atmosphere. Hydrocarbons have also been found in solid and gaseous form on a number of comets and asteroids. They have even been found in interstellar space.

It seems likely that the widespread existence of hydrocarbons throughout the solar system and the universe beyond is a matter of chemistry, not biology.[6] Why, then, do we need to invoke a biological explanation to explain the existence of hydrocarbons on our particular planet? When faced with two theories that explain a given set of facts equally well, we generally prefer the simpler of the two. If earthly hydrocarbons are the product of buried decaying vegetation, we have to invoke different explanations for the existence of hydrocarbons on Earth and on other planets; but if hydrocarbons are primordial, only one explanation is needed. However, Gold's preference for the abiogenic (primordial) explanation of the existence of earthly hydrocarbons rests on much more than the Occam's razor argument. His theory simply explains much more about the nature of our planet. Let us consider fifteen types of observations noted by Gold

[5] Methane, ethane, and other hydrocarbon gases have been detected in planetary atmospheres based on their spectra, and tar has also been detected.

[6] However, we cannot exclude the possibility that subsurface life might exist on many planets, and on their moons.

 that are difficult to understand if oil and gas have a biogenic origin. (I will deal with coal separately.)

Problems with the Biogenic Theory of Oil and Gas

1. Crude oil has the wrong chemical composition. Molecules having different arrangements of the same atoms are known as *isomers*, as in the example of right- and left-handed sugar molecules. Natural petroleum has the same subset of carbon-hydrogen isomers that are found in synthetic oil rather than in oil that is produced biologically.[6] Natural petroleum also appears to be too hydrogen-rich to be the product of debris from plankton and other marine organisms, which instead would be expected to favor unsaturated hydrocarbons.[7] Indeed, no oil which chemically resembles natural petroleum has ever been made artificially from plant material under conditions similar to those found in nature.[8] Supporters of the conventional theory note that the failure to make artificial oil that chemically resembles natural petroleum can be easily explained because organic materials maturate differently, depending on how long they are "cooked." Just think of the different results if your Thanksgiving turkey were cooked slowly in an oven or rapidly in a blast furnace. In rebuttal to Gold, they also point to biomarkers found in petroleum such as porphyrins, which are claimed to occur only as the result of photosynthesis.

2. Sediments often lack fossils. If hydrocarbons had a biogenic origin, we might expect their source to be rich in fossils, but the reverse is usually the case.[1] Oil is often found in sediments that completely lack fossils. (The paucity of fossils is explained within the conventional theory by noting that most of the material from which petroleum is formed comes from plant material and simple one-celled organisms, such as algae and bacteria, that lack the hard parts that would be preserved as fossils after being "cooked" for geological intervals of time.) Shale sometimes has fossils and is thought to be the "source rock" from which nearby oil is generated. But, according to Gold, an equally good explanation is that the shale became saturated with oil nearby after it had been formed. Moreover, some tar deposits in Canada completely lack any so-called source rocks.[8]

3. ***Deep petroleum lacks biological traces.*** The chemical composition of petroleum from very deep levels is found to lack nearly all the evidence for a biological origin seen in petroleum found in shallower sedimentary deposits.[1] If all petroleum were of biological origin, why should that from deeper levels show no optical activity and have all traces of biological molecules destroyed, while remaining chemically very similar to shallower petroleums in all other respects? This is easy to understand, however, if bacteria are simply contaminating primordial petroleum that comes up from the Earth's mantle. The petroleum brought up from deeper wells has not yet passed through the higher levels where bacteria are found. Generally, deeper hydrocarbon deposits are more hydrogen-rich, which is also to be expected under the biological contamination scenario, since bacteria would tend to deplete oils of their hydrogen content. Supporters of the conventional biogenic origin theory can argue, however, that petroleum found in deeper levels is likely to have been "cooked" for a longer geological time, which accounts for fewer biological traces.

4. ***Oil from each area has a chemical signature.*** The oil fields of the Middle East span a 2,700 kilometer stretch from

 Turkey to the Persian Gulf, where the oil is found in rocks spanning a wide range of geological ages. The times at which sediments were deposited greatly vary at different locations, and so did the prevailing climate and vegetation. Yet, strangely, oil from anywhere in the Middle East has a similar chemistry that clearly identifies it as being of Middle Eastern origin. (The same is true for oils from other regions, which each have a distinctive signature.) If local sedimentary deposits had been the origin of the petroleum, it is hard to see from the greatly variable ages of the deposits how such a common areawide signature arose. Conversely, such a common signature is easy to understand if all oil in a given region seeped up from a common source deep within the Earth's mantle.

5. Oil and gas are often found in long linear or arc-shaped regions. The distribution of hydrocarbons around the globe also argues for its abiogenic origin. Gold notes that "petroleum and methane are frequently found in geographic patterns of long lines or arcs, which are related to deep-seated large-scale structural features of the crust rather than to the smaller scale patchwork of the sedimentary deposits."[4] Thus, for example, oil and gas fields in Southeast Asia can be found along a 6,000 kilometer circular arc from China to New Guinea. This arc is also a region of much volcanic and earthquake activity. There are great differences in the age of sedimentary deposits and in many other aspects of the geology and topography of the oil fields along this arc. The circular regularity of the arc and the belt of earthquakes and volcanoes all along it both suggest to Gold that oil and gas are being produced deep in the Earth rather than in disparate sedimentary deposits laid down during different geological eras. However, supporters of the conventional theory argue that these arc-shaped regions are a feature of colliding tectonic plates and of places where sedimentary beds are deformed.

6. Hydrocarbons are found at all depths. Although petroleum is mainly found in sedimentary deposits, it is also found (along with hydrocarbon gases) in the deeper underlying rocks. If hydrocarbons were of biogenic origin, we would expect deposits to be found at various strata corresponding to different geological eras of abundant vegetation. Hydrocarbons, being lighter than the surrounding rocks, should migrate upwards but not downwards. Therefore, we would expect to find some earliest-deposited layers of hydrocarbons, below which no traces are found. In fact, however, some hydrocarbons are found at all levels below any substantial accumulation—a rule first enunciated by N. A. Kudryavtsev, perhaps the strongest modern proponent of an abiogenic origin of hydrocarbons.[8] Kudryavtsev's rule has been confirmed in many locations, including Oklahoma, Wyoming, Canada, Iran, Java, Russia, and Sumatra. In some of these cases, it could be argued that hydrocarbons were found due to the horizontal migration from other nearby oil or gas fields, or due to contamination from fluids used in the drilling process. One case where these alternative explanations can be ruled out is in a pair of test wells that Tom Gold persuaded the Swedish government to dig down to 6.7 kilometers. In one of these holes, only water-based mud was used as the drilling fluid in order to avoid any possible contamination problem. Liquid petroleum was found in cracks in the "basement rocks," with no sedimentary deposits anywhere nearby, and a total of fifteen tons of oil was pumped up.[9] The Swedish basement rocks were of the metamorphic type, meaning that they were formed when rocks of the igneous or sedimentary type were transformed by intense heat and pressure. Thus, any oil in the

131 original rocks would not have survived their melting and freezing to their present condition. Oil found in cracks in metamorphic rocks could only come from deeper layers.

One counterargument to Gold's theory is that the distribution of petroleum is very uneven, with rocks formed during certain periods of Earth's history (during which life is presumed to have flourished) containing much higher concentrations of petroleum than other periods. Gold responds, however, that it is only an assumption that life was more abundant at those times, and the reason why those rocks have greater petroleum accumulations has to do only with their relative porosity and their ability to store petroleum coming up from below.

7. Methane is found in biologically improbable places. We have already noted that methane (the primary component of natural gas) is common throughout the solar system. Even on Earth it is found in many places where a biological origin seems improbable, or where the extent of sedimentary deposits are inadequate. Such improbable places include fissures in basement rocks, deep ocean rifts, and active volcanic regions. In addition, methane hydrates (mixtures of methane and water ice) have been found by Russian scientists at every location under the ocean floor and the permafrost where the pressure and temperature conditions would permit them to exist.[10] The amounts found cannot be accounted for on the basis of biological deposition.[10] Extensive flames seen in some major volcanic eruptions are also a possible sign of natural gas being brought up from great depths.[7] The amount of methane re-

[7] During quiet times between eruptions, when bubbles of methane arise from great depths, the gas comes into contact with a much greater surface area of molten rock, and it is oxidized to CO_2 before reaching the surface.

 leased in such eruptions has been estimated to exceed even the most prolific known gas pools, which indicates a much deeper origin.[8]

8. Surface soils above gas fields have a very high methane content. The surface soil above natural gas fields is highly enriched in methane, sometimes over a hundred times the background level.[4] Such a high methane soil concentration implies a high gas flow rate from below. In some cases, methane flow rates would need to be so high that any gas field would be depleted in just a few thousand years. This makes little sense if the methane and other hydrocarbons were created in deposits laid down many millions of years ago.[4] On the other hand, the high methane soil concentrations (and the high methane flow rates they imply) are easy to understand if the gas seeps up from much larger reservoirs deep inside the Earth.

9. Helium is always found in association with methane. A particularly potent argument against the standard biological origin theory of hydrocarbons is the strong association found between helium and methane gas. Helium never comes to the surface in pure form but is always mixed with methane, in varying concentrations. The puzzling association of helium with methane is very difficult to explain if methane originates in buried sediments. Being an inert gas, helium takes part in almost no chemical reactions, so it is hard to see why chemical reactions involving decaying vegetation would tend to concentrate the gas. One known place where helium is being continually generated inside the Earth is from the alpha decay of uranium.[8] Given the slow nature of this decay process, whose half-life is about 4.5 bil-

[8] An alpha particle is the nucleus of a helium atom.

 lion years, helium accumulates very gradually in rocks. The quantities of helium that build up in rock fissures are not great enough for the gas to attain enough pressure to reach the surface on its own—hence no pure helium gas is released. However, when methane gas flows through porous rocks, it can sweep helium along with it.

Figure 7.1. Failure to shut off the flow from the world's largest underground helium well leads to catastrophe. (Cartoon is from my book, *What If You Could Unscramble an Egg?*, published by Rutgers University Press.)

Because helium is being created continuously from radioactive decay, more will be swept up by methane, the greater the depth from which methane originates. The amount of helium concentration in methane when it reaches the surface is, therefore, an indicator of the depth from which the methane originated. If methane had a biological origin in shallow sedimentary deposits, its rise to the surface would not sweep up nearly the concentration of helium that is found in many cases (up to 10 percent of natural gas). It could be argued that cases of high helium concentration in methane merely reflect local uranium deposits rather than a great depth of origin. However, this explanation suggests that high helium concentration in methane is a promising way to search for uranium deposits, which is not found to be the case. High helium concentrations in methane, therefore, imply that the methane has come up from great depth, far below the sedimentary levels, and has swept up the diffuse helium along its path.

10. Rocks at great depths can contain open pores in isolated domains. Fluids can only come up from deep underground if there are pores or cracks in the rocks. If there is not a sufficiently high fluid pressure from below, the weight of the overlying rocks tends to close any such rock openings below a certain depth, but this is not a problem if the fluid in the pores or cracks has the same density as in the surrounding rock itself. For this reason, molten lava is able to come up from the mantle at depths of perhaps 200 kilometers in the case of volcanic eruptions. The situation is different, however, in the case of lower-density hydrocarbon fluids. In this case, the weight of the overlying rock will crush rock pores closed at a critical depth of between 3 and 10 kilometers.[4]

Based on the preceding, geologists have argued that no open rock pores (and hence no deep reservoirs of hydrocarbons) could exist below some critical depth.[11] But, according to Gold, this conclusion is fallacious. The proper conclusion is that at a critical depth of 3 to 10 kilometers, closed rock pores create a barrier. That barrier allows the pressure of any fluid below to build up enough to keep the pores open for another 3 to 10 kilometers below the barrier, and so forth for any number of lower domains. In other words, reservoirs of methane in porous rocks deep underground should be found in stacked domains of vertical heights between 3 and 10 kilometers. Closed rock pores at the bottom of one domain tend to prevent the gas from flowing upward into the next higher domain. As we descend through the Earth, we should therefore find that the fluid pressure just below a domain boundary rises abruptly because closed pores prevent upward fluid migration. So much for the theory. But what does the drilling record actually show? Most gas drilling has been confined to depths corresponding to the topmost domain. But there have been some cases of deep drilling during which a sudden increase in gas pressure was encountered, indicating a breakthrough to a lower domain.[4] This observation suggests that methane may indeed be found in a succession of vertically stacked domains extending far underground.

11. Petroleum reservoirs refill spontaneously. In many cases it has been noted that petroleum reservoirs seem to refill themselves partially as oil is pumped out of them, a phenomenon that has been particularly noted in oil fields in the Middle East and in the U.S. Gulf Coast. As a result, original estimates of how long oil supplies will last often have turned out to be gross underestimates. Gold's theory of deep sources of hydrocarbons nicely accounts for the spontaneous refilling of oil reservoirs from below. When oil is pumped out of the topmost domain, the pressure difference across the domain boundary is increased, and oil in the lower domain at high pressure can then force its way through the boundary and refill the upper domain.

12. Diamonds exist. Diamonds are pure carbon, just like the "lead" (actually graphite) in your pencil—but with a very different crystal structure that can be created only under conditions of extremely high pressure. Gold suggests that the presence of diamonds on Earth implies the presence of pure carbon at depths between 150 and 300 kilometers—the only place the pressure is great enough to form them.[12] But how did the carbon get to be in its pure form, separated from the other elements? Gold's answer is that the dissociation of carbon is similar to the mineralization process by which nuggets of pure gold and other minerals are created. Carbon dissociation can proceed only when carbon-bearing fluids flow through pores in rocks where other elements can be selectively removed in chemical reactions. (Without a flowing fluid, chemical reactions take place only on a solid surface, and not throughout the volume of a material.)

What could the carbon-bearing fluid be that allows regions of pure carbon to form? Traces of carbon dioxide, as well as methane and other hydrocarbons, have been found in some diamonds, so the carbon-bearing fluid could be either CO_2 or a hydrocarbon. However, the latter seems to be the likelier possibility, since hydrocarbon molecules can be dissociated more easily (at lower temperatures) than CO_2 molecules. Thus, the existence of diamonds implies that hydrocarbons flow through rocks deep underground. Diamonds formed deep inside the Earth will only survive in that form if they are cooled very quickly, otherwise they change to graphite. Thus, diamonds are found on the surface often as a result of explosive gas eruptions occurring in funnel-shaped depressions known as diamond pipes. Hydrocarbons found in these pipes give further evidence for the role of hydrocarbons in forming diamonds, and therefore for the presence of these substances deep inside the Earth.

13. The Earth's surface layers are very rich in carbon. The Earth's surface and the sediments are extremely carbon rich, most of the carbon being in the form of carbonate rocks. The total amount of carbon on the surface and in the sediments amounts to perhaps 200,000 kilograms for each square

137 meter of surface area. The high concentration of carbon in Earth's surface layers can be explained by Gold's theory, while the conventional biogenic origin theory of hydrocarbon is silent on the matter. Let's consider the possible origin of the carbon on the surface and in the sediments. One possibility is that the high carbon content is due to the recycling of carbon (i.e., the "carbon cycle"), but Gold argues that both the amounts involved and the isotopic composition of the carbon make this highly unlikely.

Another possibility is that the Earth was originally formed with a superdense atmosphere having a very large amount of carbon dioxide, which was later laid down as carbonate rocks. However, in that case, the original atmosphere would also need to have had substantial amounts of inert gases as well as CO_2, because a constant relative abundance of the elements is found in the interstellar material from which the Earth was formed. Having no significant chemical reactions, any neon, argon, and krypton in the original atmosphere would remain to the present day. Based on the relative scarcity of these inert gases in the present atmosphere, we can conclude that the Earth did not have an initial superdense atmosphere with a lot of CO_2. Another possible origin of the Earth's abundant surface carbon is the infall of carbon-rich meteorites. But, this possibility can also be ruled out, based on the cratering record on Earth (and also on the erosion-free moon), which would have to be more extensive in order to account for the large amount of carbon found on Earth.

Meteorites may, however, successfully account for the Earth's surface carbon abundance in another way. The Earth is believed to have formed from the chunks of rock that are essentially the same as the meteorites that still hit the planet today. One particular type of meteorite, known as carbonaceous chondrite, is very carbon rich. Carbonaceaus chondrites may comprise the most common type of meteorite, but this is uncertain, since the meteorites are difficult to spot on the ground; they also tend to disintegrate more easily than other types, both before and after impact. When these meteorites are heated under high pressure, they release hydrocarbon fluids. If a substantial fraction of the Earth were formed from coalescing rocks of the chondrite type, we would expect that substantial amounts of hydrocarbons were released deep inside the Earth under conditions of high pressure and temperature. Such hydrocarbon fluids, created at depths of perhaps 200 kilometers, would be driven upwards due to their buoyancy and could explain the carbon richness of the outer portion of the Earth's crust and sediments.

14. Carbon isotope fractionation in methane varies with depth. The element carbon has two stable isotopes, C_{12} and C_{13}, whose relative abundances are roughly in the ratio 100 to 1. Chemical and physical processes generally favor one isotope over the other by only a very small degree. During the process of photosynthesis, for example, plants using CO_2 tend to slightly favor the lighter carbon isotope C_{12}. Therefore, most geologists take the presence of hydrocarbons with enriched levels of C_{12} (or depleted levels of C_{13}) to be a sign that biological processes were involved in hydrocarbon creation. It has been generally assumed that any material that contains carbon C_{13} depleted by more than 2 percent below its "normal" value has a biological origin.[13] (In other words, if the normal level of C_{13} were exactly one percent, the test of biogenic origin would correspond to a C_{13} level of 0.98 percent or less.) But it is misleading to imagine a sharp dividing line between biogenic and nonbiogenic carbon based on their C_{13} depletion level. In fact, no such

139 clear distinction exists. The carbon found in carbonaceous chondrite meteorites, for example, is found to have C_{13} depletions of anything from –2 to +11 percent.[4] If the 2 percent depletion rule were taken literally, we would have to conclude that a substantial fraction of chondrite meteorites have been involved in biological processes *in space*.

The data on C_{13} depletion in methane, which has been said to support a biogenic origin of hydrocarbons, actually better supports Gold's abiogenic theory. Data on the extent of C_{13} enrichment in methane gas as a function of depth shows that the gas has less C_{13} the closer it gets to the surface.[14] Why should this be so? When a gas diffuses through a porous material like rock, molecules with the lighter carbon isotope C_{12} move slightly faster and move through the rock pores more quickly. The greater the amount of rock through which the gas passes, the greater the depletion of C_{13}. (The same process of gas diffusion is used commercially to separate uranium isotopes to make enriched nuclear fuel.) In the case of methane, other processes besides gas diffusion, including reduced rates of absorption and oxidation for C_{13} compared to C_{12}, also act to enhance the extent of C_{13} depletion, as the methane rises.

Three specific observations of the pattern of carbon enrichment in methane match Gold's theory of deep underground methane sources: (a) the range of C_{13} depletion levels found in gaseous methane is ten times that for petroleum, indicating that the cause is gaseous diffusion rather than biology (which would predict similar ranges for the two hydrocarbons); (b) methane from deeper levels has less C_{13} depletion, since less has been removed by diffusion and other processes; (c) methane from even the deepest gas fields explored to date show significant C_{13} depletion (since the methane sources are far deeper).[15]

 15. Carbon isotope fractionation in marine carbonates is constant. Carbonate rocks are laid down when CO_2 is extracted from the atmosphere. Recall that plants favor the light C_{12} isotope in photosynthesis. When vegetation dies and is buried in sediments, proportionately more C_{12} than C_{13} is removed from the atmosphere. As a result, we would expect that carbonate rocks laid down from atmospheric CO_2 during geologic times when vegetation flourished should show higher than average levels of C_{13}. These variations in C_{13} enrichment should be readily observable in carbonate rocks if as much as a fifth of the buried carbon were the result of decaying vegetation.[16] In fact, however, carbonate rocks show very little variation in the proportion of C_{13}, even during the Silurian time when vegetation suddenly proliferated on Earth.[16] The only plausible explanation is that most of the hydrocarbon deposits are not due to buried vegetation, but instead have an abiogenic origin.

What About Coal?

In the standard biogenic origin theory, coal results from the product of buried decaying vegetation, in contrast to oil, which is believed to have resulted from the remains of marine organisms. At certain wet and warm times in the Earth's history, vast stretches of swamp forests proliferated. When the decaying vegetation was submerged in oxygen-depleted swampy water and was overlayed with a sufficient layer of rock and soil, the high pressure and temperature converted the material into coal over the course of hundreds of millions of years—at least that's the generally accepted view.

Most geologists are willing to admit that there may be deep abiotic sources of methane, but they are skeptical of

 Gold's theory that all oil and gas have an abiogenic origin, and they are completely incredulous about his contention that coal—at least black coal—also has an abiogenic origin. Unlike the case of oil and gas, for example, coal deposits usually (but not always) do contain the organic remains of plants, which would seem to make the case for a biogenic origin fairly clear. If hydrocarbons get their biological traces merely due to deep underground bacteria, how can we explain the presence of spores, wood, and degraded particulate matter from plants often found in coal?

Gold's answer is that upwelling hydrocarbon fluids from the deep crust and mantle often encounter sedimentary deposits, where the fluid is gradually turned into the substance we call coal, and the fossils in those sediments become embedded in the coal. He suggests the process is quite analogous to the process by which petrified wood is formed. Gold presents a number of different lines of evidence, relating to the composition and distribution of coal, in support of this view. He suggests that even the presence of fossils offers stronger evidence for his theory than the standard biogenic origin.

When fossils are found in coal, they are usually highly compressed, but sometimes they are entirely filled with carbon without being deformed, even at the cellular level. This observation suggests to Gold that a liquid hydrocarbon initially filled the fossils; the finding is not consistent with the surrounding body of coal being formed from the fossilized material. Moreover, sometimes an intact leaf or twig is found in coal. If the coal was formed from vegetation, how could the process leave some material completely intact while converting the surrounding material into a homogenized mass?

Biological fingerprints are also present in coal at the molecular level, just as they are in petroleum. Here, too, Gold argues that the facts better support his theory than the standard one. The particular molecules found in coal that point to signs of biological activity are the same ones that are found in petroleum. That seems like a strange coincidence, if coal is the result of decaying remains of land vegetation, while petroleum results from marine life, but it is just what Gold's theory would predict.

Moreover, the mineral content of some black coals is too low to agree with the hypothesis that coal formed from decaying vegetation found on the planet today—although, of course, the climate and vegetation might have been very different when the sediments were formed. The content of many rare metals, including uranium, gallium, and mercury, in coal is also too high to be explained by the conventional theory. On the other hand, if coal is the result of upwelling hydrocarbon fluids, such minerals and metals could be leached out of the surrounding rocks as the hydrocarbon fluid traveled upwards through rock pores.

The frequent geographic association of coal with methane and petroleum is also difficult to understand if coal is the result of the remains of land vegetation while petroleum is the product of marine life. But it is easily explained in Gold's theory. As upwelling hydrocarbon fluids rise through the Earth, they progressively become more carbon enriched and hydrogen depleted through chemical reactions with stray oxygen atoms. The result is that coal seams would be formed above gas and oil fields, as is often found to be the case.

In Gold's theory, coal is formed when hydrocarbons flow upwards through porous rocks. On occasion, we might expect the result to be coal deposited in vertical seams—which, in fact, is the case. Such near-vertical seams are difficult to explain by the conventional theory of coal forming from buried sediments. However, the absence of hydrocarbons in high porosity sands below coal deposits is certainly difficult to reconcile with Gold's theory of coal formation. Gold does acknowledge that "brown coal"—peat and lignite—offers clear evidence of a biogenic origin. In one sense, this acknowledgment weakens Gold's case, because if one type of coal has a biogenic origin, it seems simplest to argue that all types should. However, Gold argues that it is only a matter of semantics that peat and lignite are even classified as "coal," because there is no continuous gradation from them to black coals.

How Would We Know if Gold's Theory Is Correct?

Thomas Gold may now be in a small minority who question the biogenic origin of hydrocarbons, but the idea dates back to the nineteenth century. Early proponents of a nonbiogenic origin include such luminaries as the chemist Dmitry Mendeleyev, who discovered the periodic table of the elements. At the beginning of the twentieth century, early estimates of the Earth's total amount of oil and gas were less than one percent of the present estimate. According to Gold, our present estimates could be off by at least as large a factor as those earlier estimates. If Gold is correct, the Earth is likely to have enormously greater (hundredfold or more) hydrocarbon reserves than are currently believed. These reserves would be deeper than the reserves so far explored and would be more expensive to reach, but the extra expense of drilling might be offset by the greater abundance found, since natural gas from deeper levels is under much higher pressure and has greater density than gas from shallower levels.[9] Recent deep-well gas prospecting has had a very high (70 percent) success rate, which is what Gold's theory would predict if deep gas is ubiquitous.

According to Gold's theory, nonsedimentary rocks (largely avoided by petroleum geologists) should be just as promising a place to drill for oil as sedimentary rocks, provided they have sufficient porosity. To the extent that judgments of where to drill for petroleum are made by geologists who are extremely skeptical of Gold's theory, it could be argued that current statistics of oil discovery and exploration shed little light on the validity of Gold's theory. On the other hand, if new deep gas and oil discoveries should be made in nonsedimentary rocks at farflung areas of the globe, Gold's idea would become increasingly plausible. His theory also implies that prospecting for natural gas based on soil concentration of methane could prove quite fruitful—a practice that has not been widespread, supposedly because of doubts that such soil concentrations could really be the result of gas seepage from great depth.

If Gold is right, the stakes for humanity are enormous, which makes it all the more important to examine his hypothesis with an open mind, unburdened by long-held beliefs whose basis does not rely on fundamental principles. Currently, the world relies on the so-called fossil fuels for a major portion of its energy. If humanity is to have a long-term future, eventually we will need to switch to renewable options. Abundant sources of coal, oil, and gas would make that transition much easier. It would also reduce the chances of global conflict that might arise from uneven access to energy supplies. On the other hand, superabundant supplies of coal, oil, and gas would be a mixed blessing. While the

[9] At a depth of 6 kilometers, methane would be four hundred times as dense as on the Earth's surface.

 extent of climate change resulting from the burning of these fuels for another century might be tolerable (or possibly even beneficial), their indefinite long-term use is bound to lead to problems.

You might imagine that, given the financial stakes involved, some oil companies would have taken advantage of Gold's theory (assuming he were right), and their lack of interest, therefore, would seem to argue against the correctness of his theory. But it is dangerous to make arguments based on the motivations of oil company executives. One could argue, for example, that it is in the interest of the oil companies to keep oil prices high by promoting an image of scarcity, and that currently they would have little interest in finding that oil is far more plentiful than had been thought.

My rating for the idea that coal, oil, and gas do not have a biogenic origin, and were part of the Earth's original composition, is zero cuckoos.

8 Time Travel Is Possible

In our memories, dreams, and imagination, we regularly "travel" to the past and future. But what if you could actually hop into a machine such as the one in H. G. Wells's famous science fiction story, *The Time Machine*, which would take you to any desired past or future time? Where would you want to go? If time travel became commonplace, perhaps travel agents would be busy booking trips to important historical events, which would then become packed with time travelers from *all* future times. But how could that possibly be true? Certainly, if hordes of time travelers were present at such events, their presence has not been noted by historians. In fact, the very absence of such reports of hordes of time travelers has been suggested seriously as an argument against the possibility of time travel. This argument (among many others) leads many people to regard time travel as belonging strictly to the realm of science fiction. Surprisingly, some physicists do not share this view.

Spacetime Diagrams and Worldlines

In a trivial sense, all of us are time travelers right now—moving forward into the future at a rate of one second per second. But meaningful time travel involves changing, and possibly reversing, that progress through time, or perhaps jumping across time. The physics of time travel is closely connected with Einstein's theory of relativity, in which 3D space and 1D time are linked into an integrated four-dimen-

147 sional "spacetime." Any specific event which occurred at a particular place in space and time, such as your last breath, can be represented as a point in a spacetime (Minkowski) diagram. Spacetime diagrams can also be used to display the movement of objects through space and time (see fig. 8.1). Note that without being able to draw a four-dimen-

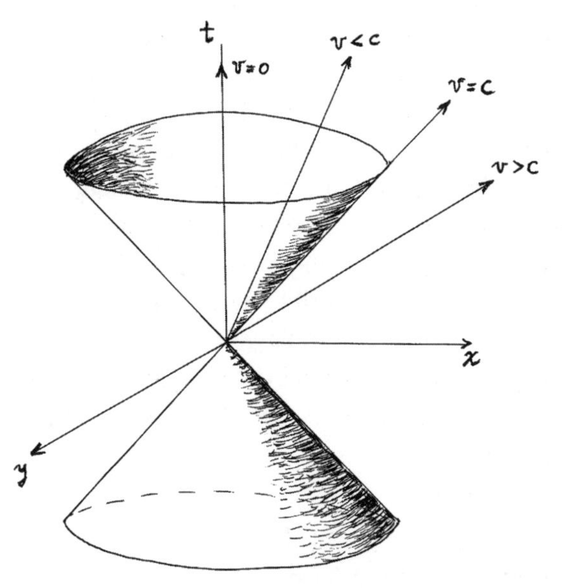

Figure 8.1. A spacetime diagram showing the past and future light cones and worldlines for objects having four different velocities. In appropriate units for *x* and *t*, the worldline of an object traveling at the speed of light *c* makes a 45 degree angle with the time axis, and it lies on the surface of the light cone.

sional figure, we can show only two of the three dimensions of space, *x* and *y*, and the one time dimension, *t*, along the vertical direction. A line showing the progress of an object through spacetime is known as its "worldline." Assuming you are stationary ($v = 0$), your "motion" in this diagram would be only along the time dimension, and your worldline would be vertical.

Worldlines of moving objects (which travel through space as well as time) are tilted, with the angle of tilt away from the vertical being greater, the faster the object moves. Suppose we measure time in years and distance in light years. In that case, the worldline for a light ray (which travels one light year distance in a time of one year) would make a 45 degree angle with the time axis. Light rays traveling in different directions starting from the origin must all lie on the surface of a 45 degree "light cone," shown in figure 8.1.

If an object travels at speeds less (greater) than that of light, its worldline lies entirely inside (outside) the light cone, assuming that the worldline started at the origin. Worldlines lying outside the light cone are said to be "spacelike" (since they move mostly through space), while worldlines entirely within the light cone are said to have "timelike" worldlines. Worldlines need not be straight lines, which would describe objects that travel at constant speed in a straight line. Can you imagine what the worldline of a satellite traveling in a circular orbit would look like? HINT: Imagine what path you would follow in three dimensions if you traveled around a horizontal circle while at the same time steadily advancing upwards along the time dimension (see answer in footnote).[1]

Regions of Spacetime and the Light Cone

The two light cones shown in figure 8.1 divide spacetime into three distinct regions. The "future" is the region of spacetime inside the upper light cone, the "past" is the region inside the lower light cone, and "elsewhere" ("elsewhen" is just as appropriate) is all the rest of spacetime outside the cones. What is the basis for this strange threefold division of spacetime?

[1] It would be a helix with its axis along the vertical time dimension.

Before relativity came along and messed up our commonsense view of space and time, the two were regarded as being independent entities. Let's assume that the origin of spacetime (the point where the two cones meet) represents an event occurring right here and now ($t = 0$). In classical physics with its absolute time, all observers would agree whether an event was in our future ($t > 0$) or past ($t < 0$). But relativity does away with absolute time. Using the equations of special relativity (the Lorentz transformation), we find that the space and time coordinates of an event depend on one another.[2] As a result, for us, successive instants of time correspond to the series of equally spaced stacked horizontal planes in figure 8.1, but a moving observer would find that succesive instants of time correspond to a series of stacked, tilted planes. Despite the tilt of the time planes for a moving observer, we find that any point we consider to be inside the future (upper) light cone will remain there no matter what the motion of the observer. Likewise, any relatively moving observer will find that points in the past (lower) light cone stay there, no matter what his state of motion. In this sense, points in the upper (lower) light cone are in the *absolute* future (past).

What of points in the "elsewhere" region that falls outside the two light cones? In this case *you* may judge the event to have $t > 0$, but an alien moving past at sufficiently high speed could find that $t < 0$. This change in sign of the time amounts to switching an event from the future to the past. Scrambling the time order of events is an essential feature of relativity in which space and time are interlocked. Might this kind of switching of events from future to past offer a

[2] The Lorentz transformation gives the x' and t' coordinates of an event in terms of x and t, where $x' = \gamma(x - \beta t)$ and $t' = \gamma(t - \beta x)$, where $\beta = v/c$ and $\gamma = 1/\sqrt{1 - \beta^2}$.

 means of time travel? Certainly not, unless we could send objects with faster-than-light speeds ($v > c$), which corresponded to spacelike worldlines (outside the light cone)—a subject discussed in the next chapter.

One-Way Time Travel into the Future

One form of time travel which nearly all physicists agree *is* possible involves one-way travel into the future, based on the time dilation effect. In the well-known "twin paradox," one twin, Lisa, stays behind on Earth, while her brother, Bart, hops aboard a spaceship on a round trip journey to a distant star. When Bart returns to Earth he would be younger than Lisa, with the difference in their ages depending on the speed of Bart's spaceship during the journey. The ratio of Bart's elapsed time to that of Lisa's becomes smaller, the closer his speed approaches the speed of light.

If Bart were to travel at 98 percent light speed, for example, one year would elapse for Bart for every five years for Lisa.[3] Suppose Bart journeys to a star ten light years away. Lisa would judge that Bart's round-trip took slightly more than twenty years (Earth time) to cover the round-trip distance of twenty light years. But, Bart would find that the trip lasted only slightly more than a fifth as long, or four years. Therefore, Bart travels just over twenty years into Earth's future, while aging only four years in the process.

This twin paradox example is illustrated in the spacetime

[3] The ratio of Bart's to Lisa's times is given by the factor $\sqrt{1-\beta^2}$, where $\beta = v/c$ is Bart's speed expressed as a fraction of the speed of light. If Bart travels at 98 percent the speed of light, or $\beta = 0.98$, we find $\sqrt{1-0.98^2} = 1/5$.

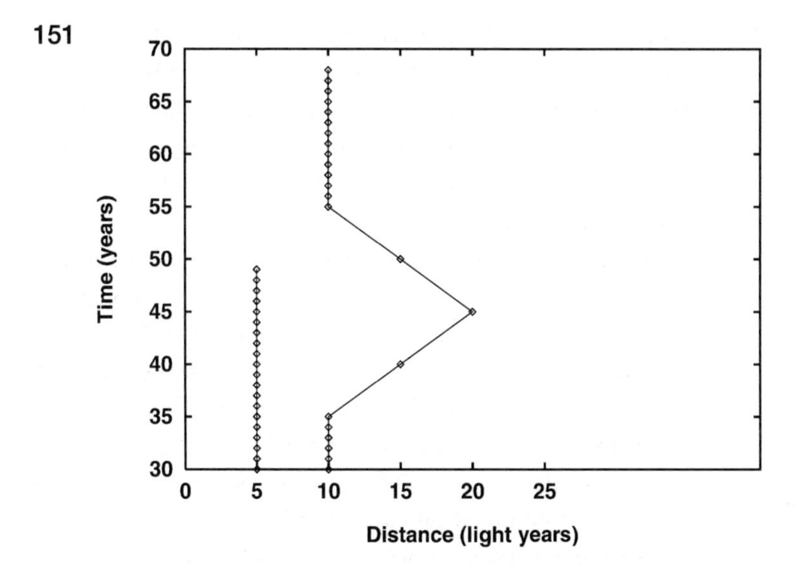

Figure 8.2. The spacetime representation of the "twin paradox," showing portions of the worldlines for Lisa and Bart, a pair of twins. Lisa, the stay-at-home twin, has the vertical worldline. The twenty-three dots on each twin's worldline show successive birthdays. Bart, the space-traveling twin experiences four birthdays during his interstellar journey, while Lisa experiences twenty. Thus, Bart, after he returns home, doesn't reach his *n*th birthday until sixteen years later than Lisa.

diagram shown in figure 8.2, where we show distance from Earth (in light years) on the horizontal axis, and time (in years) on the vertical axis. For clarity we have separated Bart and Lisa's worldlines—Lisa's being the vertical line at five light years, and Bart's being the bent line starting at ten light years. Lisa's worldline is vertical because she "moves" only through time. Each twin's birthday is shown as a point on his or her worldline. Bart spends his first thirty-five years on Earth, then travels at nearly the speed of light, reaching a star about ten light years away roughly ten years later (according to Lisa), and returns to Earth when Lisa would have been fifty-five. Note that Bart only has four birthdays during his trip, which lasted twenty years in Earth time. The abrupt change in slope of Bart's worldline at the star would correspond to a sudden reversal in his velocity, which is unrealistic in terms of the crushing impact such a reversal would have on Bart.

The differential aging of the twins has been called a *paradox* because based on the relativity of motion, we might suppose that Bart could just as well regard himself to be the stationary stay-at-home twin, and Lisa the traveling twin. In that case, Lisa rather than Bart should be the one whose time runs slow. But it is only *uniform* motion that is relative. Bart knows that he was the one who took the journey because of the accelerations and decelerations that he could feel, so there is no symmetry between the two twins, and Bart really does age less than Lisa.

In principle, such one-way time travel could take place arbitrarily far into the future if it were possible to achieve the necessary speed. For example, if you wanted to go one million years into the future (Earth time) on a trip that lasted according to you only one year, you could do so if your speed were 99.99999999995 percent the speed of light the whole way.[4] Of course, you couldn't achieve extreme speeds very rapidly, given the inability of humans to withstand large accelerations. Nevertheless, speeds extremely close to that of light could be obtained by accelerating a spaceship for a modest number of years at a constant 1 g acceleration like that prevailing on Earth.

In such a case, it can be shown that the time elapsed on Earth (in years) would be given by the natural logarithm of

[4] See previous footnote for the formula, but you probably won't be able to check this result on your calculator, because most calculators can only handle numbers with fewer than thirteen digits. But there is a simple rule

153 twice the time elapsed according to someone in the spaceship.[1] For example, for one million years of Earth time to elapse, the spaceship would need to travel for 14.5 years according to its own clocks with a one g acceleration or deceleration. Of course, the technical difficulties associated with such trips might make them forever impractical, but the fact that they are theoretically possible is immensely interesting. On the other hand, without the possibility of a return trip to the past, you might consider this kind of time travel more akin to suspended or slowed-down animation rather than true time travel. But there is an important difference: in the example just considered, your ship would only need 14.5 (not one million) years worth of fuel.

The physics behind the twin paradox has been verified experimentally. Unstable subatomic particles have been shown to have their lifetime lengthened when they travel at very high speeds. For example, a particle traveling at 98 percent the speed of light (like Bart) travels five times further before disintegrating than it would if we ignored time dilation. Similar effects of a much smaller magnitude have been found for speeds in the domain of human experience. Ultrasensitive clocks taken on board a jet aircraft and flown around the world have been found to run slow by a very slight amount of time compared to clocks that stay on the ground.[5] Presumably, humans on board the aircraft age less by the same amount as the electronic clock's time runs slow, but such minute effects on human body "clocks" could not possibly be measured.

---

for taking the square of a number very close to one: the square root of $1 - x$ is approximately $1 - x/2$.

[5] In fact, the direction of travel also matters. Planes that fly eastward (in the direction of the Earth's rotation) have a greater velocity, and show a greater time dilation effect, than planes that fly westward.

Two-Way Time Travel

Time travel into the future might have much greater appeal if you could return to the present—perhaps with useful information that could improve your financial prospects or your love life. Most people, including most physicists, are more skeptical about the possibility of this sort of time travel. If time travel into the past or, alternatively, two-way time travel into the future were possible, all sorts of paradoxes would be created. The best known of these paradoxes involves you going back in time and killing your grandfather before he met your grandmother. By killing grandpa, your action would prevent your own future existence—so how could you have traveled back in time to perpetuate the dastardly deed?

A variety of solutions have been suggested to the killing-your-grandfather paradox, which we'll consider after we look at the physics of building a time machine. Yes—believe it or not, some physicists have described such things. But the time machines discussed by physicists differ in one important respect from that of H. G. Wells's time traveler. In Wells's story, the time traveler gets into a device that stays at one point in space and moves only through time. Such a device when traveling into the past would immediately collide with its earlier self with disastrous consequences, since two objects cannot occupy the same space at the same time. In contrast, the time travel discussed by some physicists is done using an "ordinary" vehicle that travels through a spacetime that has been distorted to make time travel into the past possible. Thus, the time machine is whatever accomplishes the distortion, rather than the actual navigation through spacetime.

155 Let's consider a possible worldline that might describe a time traveler going from A to D (at an earlier time) on a spacetime graph (see fig. 8.3). If the worldline finds itself in a flat (undistorted) region of spacetime, this worldline is impossible for a vehicle to follow. The only segments of the worldline that would be traversable are the two timelike

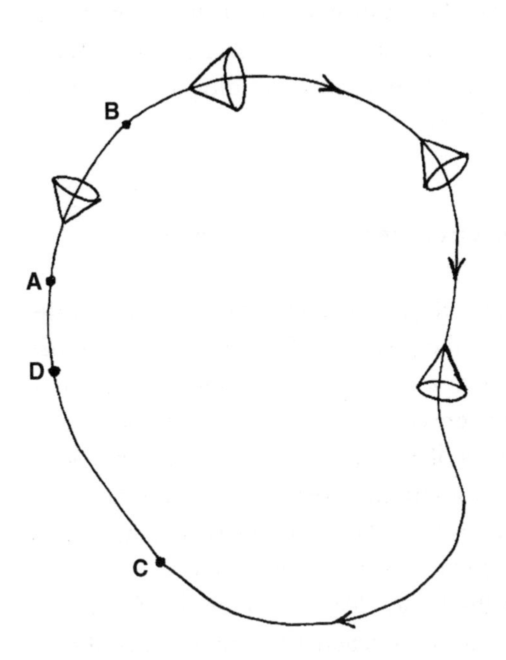

Figure 8.3. An example of a "closed timelike curve," with light cones tilted along the direction of travel. If you could follow this path through spacetime from point A to point D, you would have traveled backward in time, even though at each point on the path you would seem to be moving forward through time.

segments A to B and C to D (which have slopes exceeding 45 degrees and head into the future rather than the past). But now suppose spacetime could be distorted in such a way that the light cones at each point along the worldline were all oriented with their axes directed along the worldline. The entire closed loop D → A → D would form a *closed timelike curve.* For such a curve, a traveler following the worldline would always be moving into his *local* future (defined by the orientation of the tipped light cones at each point). Yet, when the traveler starting at point A reaches point D, he would have traveled backward in time relative to his starting point. The effect is much like that famous M. C. Escher drawing in which water flows around a paradoxical trough flowing downhill at every point yet somehow winding up above its starting point (fig. 8.4).

The question then becomes whether such spacetime distortions are actually possible, or is there no more hope of finding such a "time machine" than building a water trough of the type in the Escher drawing?

Distortion of Spacetime According to General Relativity

In Einstein's theory of general relativity, spacetime is distorted or curved by the presence of matter—the more matter density there is at a given point, the more distortion is there. In this view, gravity is considered *not* to be a force, but instead the effect of a distorted spacetime on an object's motion. In a succinct summary of general relativity, physicist John Wheeler says that "matter tells spacetime how to curve, and spacetime tells matter how to move."

To understand the basic idea, it may be helpful to consider the analogy of a stretched (flat) rubber membrane. Imagine a small weight placed at the center of the membrane which creates a depression there. Balls rolled on the distorted membrane move in ways that depend on their speed and the extent of the distortion at each point. For the right choice of speed, you could even get balls to roll in orbits around the weight causing the depression, much as planets orbit the sun. But the analogy is incomplete because it shows the curvature of space only, and it doesn't include that of the time dimension. (The distortion of the time dimension is responsible for the "gravitational redshift,"

Figure 8.4. M. C. Escher's *Waterfall* © 2000 Cordon Art B.V.—Baarn—Holland. All rights reserved. Look at the bottom of the water trough and follow the water around the trough. At each point, the water seems to be flowing downward, yet at the end of the trough it is higher than its starting point.

 which is a decrease in the frequency of light emitted from a massive body, as seen by a distant observer.)

Exactly how a distorted spacetime affects an object's motion depends on how fast it moves. For slowly moving objects (whose timelike worldlines move mostly through time), it is the curvature of the *time* dimension that explains their motion. In contrast, for a light ray which travels at 45 degrees in spacetime (and moves through space and time equally), its path is affected by the curvatures of both space and time. General relativity, in fact, predicts a very slight bending of light rays from distant stars as the light rays graze the edge of the sun—the place in our solar system where spacetime has its greatest distortion. The predicted amount of bending of a sun-grazing light ray is indeed tiny (only about 1/2000 of a degree), because even a mass as large as the sun does not distort spacetime by an appreciable amount. However, it is a large enough effect to be measured. The observation first made in 1919 that stars near the edge of the sun appear to shift in their positions by the predicted 1/2000 of a degree during a solar eclipse (the only time they would be visible) was a great verification of Einstein's theory.

Figure 8.5 depicts what the curved spacetime might look like in the vicinity of a sun much more massive than ours. Light cones are all oriented with their axes along the time direction (perpendicular to the surface shown), which is not the same at each point in space. In fact, the light cones increasingly tip toward the massive sun, the closer they are to it. When light cones tip toward the sun it means that light from the sun leaves at a reduced speed according to a distant observer. But a local observer (on the sun!) would see no change in light speed whatsoever.

If it were possible to compress the mass of our own sun into a sphere about 3 kilometers in radius, it would become a black hole.[6] In this case, the high mass density would cause such extreme distortions in spacetime near the compressed sun that nothing, not even light, could escape it. At some distance from the black hole, the light cones are tipped so much that no light ray drawn on their surface moves radially outward. This statement applies to the light cone drawn closest to the hole (or any still closer) in figure 8.5. The most extreme light ray (along the faint dotted line) travels along the vertical time direction and would never reach a distant observer. You could say that the light moves through a spacetime that is being dragged into the hole as fast as the light is traveling through it, and therefore the light gets nowhere—much like someone running on a treadmill.

What does all this discussion of black holes have to do with time travel? The connection becomes clearer when we look at rotating black holes. According to the equations of

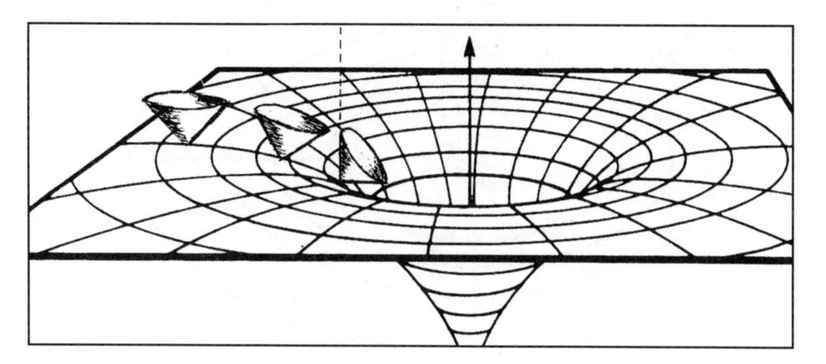

Figure 8.5. Spacetime diagram showing the distortion of spacetime in the vicinity of a black hole. The three light cones on this surface become increasingly tipped the closer they are to the center of the black hole. The closest one to the black hole is tipped so far over that no worldline on its surface moves outward from the hole.

[6] In general, a mass m would become a black hole if it were compressed to a radius r given by $2\ Gm/c^2$. Only a star much more massive than our sun could become a black hole toward the end of its life.

 general relativity, a rotating black hole literally drags spacetime around with it in the vicinity of the black hole. This dragging of spacetime can be represented by showing light cones tipped over, as in figure 8.6—not just radially inward, but also around a circle. If the rotation of the black hole is sufficiently rapid, then at some radial distance *R* the light cones are tipped all the way on their sides. Such a distortion of spacetime would *in principle* allow travel into the past. To see this, consider the spiral worldline shown in figure 8.6.

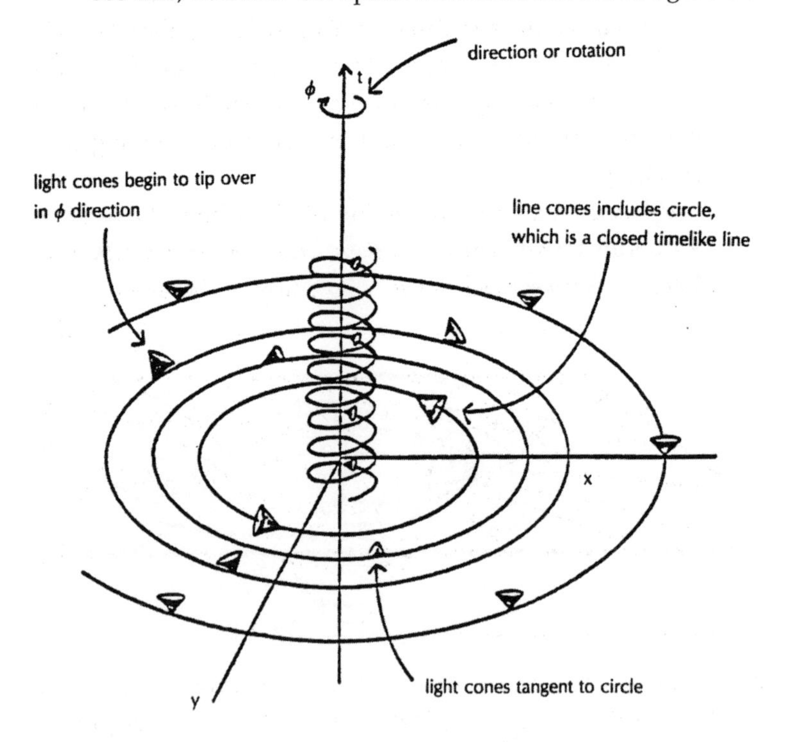

Figure 8.6. Frank Tipler's "time machine" from his 1976 Ph.D. thesis (see endnote [2]). The machine consists of a superdense rotating cylinder that deforms spacetime in its vicinity. Because the rotating cylinder warps spacetime sufficiently to put light cones on their sides, someone traveling in a circle around the cylinder (along the helical worldline) moves locally into the future but travels into the past according to a distant observer.

 (This spiral worldline corresponds to someone traveling in a circle around the black hole.)

Just as in the case of figure 8.3, a traveler following this spiral worldline in the downward direction would always be moving forward into his local future (based on the light cones). However, at the end of each loop the traveler would be at an earlier time than at the beginning of the loop. Unfortunately, time travel using rotating black holes is not possible, because the closed timelike curves circling them are inside the black hole's "event horizon." (The event horizon is the distance from the black hole from which not even light can escape.) The drawing in figure 8.6 actually first appeared not in connection with a black hole time machine, but one proposed by physicist Frank Tipler.[2] The figure, taken from his 1976 Ph.D. thesis, describes a time machine consisting of an infinitely long cylinder rotating about the vertical axis. The cylinder is a time machine because of the closed timelike curve it creates in spacetime—the innermost circle in figure 8.6. (Tipler's cylinder can't be shown in the figure because its axis lies perpendicular to the horizontal (x,y) plane, and this direction has been used to show the time dimension rather than the third space dimension z. Unfortunately, we can't draw a four-dimensional diagram!)

Tipler's cylinder would need to be made of superdense material and rotate so fast that a point on its surface would travel at least half the speed of light. Although Tipler's cylinder was infinitely long, it was later suggested that a long but finite cylinder might also work. A serious problem concerns the needed density of the material making up the cylinder. Tipler has estimated that the material might need to be so dense that a volume of the material that was the size of a single proton would contain the entire mass of the universe! Clearly, such a cylinder could not be constructed with the matter available in our universe. As with the black-hole time machine, instability is another fatal problem. To imagine Tipler's cylinder remaining stable under the immensely powerful mutual gravitational attraction between its parts is like imagining a tightly stretched rubber band remaining stretched when you released the ends. The prospects of building such a fantastic cylinder are virtually nil.

What about the idea of finding a ready-made version in nature? One possibility was suggested by Kurt Gödel's 1949 discovery that in a rotating infinite universe, the solutions to Einstein's field equations yield closed timelike curves.[3] Unfortunately, we do not live in Gödel's universe, which would need to be infinite in size, nonexpanding, and rotating faster than once every 70 billion years. We don't know if our universe is rotating, but if it is, the rotation rate is at least a thousand times slower.

Another possible ready-made time machine suggested by Richard Gott in 1991 involves cosmic strings.[4] These are extremely massive collapsed filament-like structures that some theorists believe may have been created shortly after the big bang. If cosmic strings exist, and if a pair of nonrotating strings approached each other, Gott found that a closed timelike curve would be created about the time of their closest approach. Unfortunately, Gott's solution for generating closed timelike curves appears to be just as impossible as Tipler's and Gödel's ideas. The cosmic strings involved would need to have a mass equal to most of the mass of the universe, and the closed timelike curves surrounding them would occupy an infinitely large region of space.

Wormholes

Of all the time machine proposals suggested to date, perhaps wormholes are the least crazy, although we do not

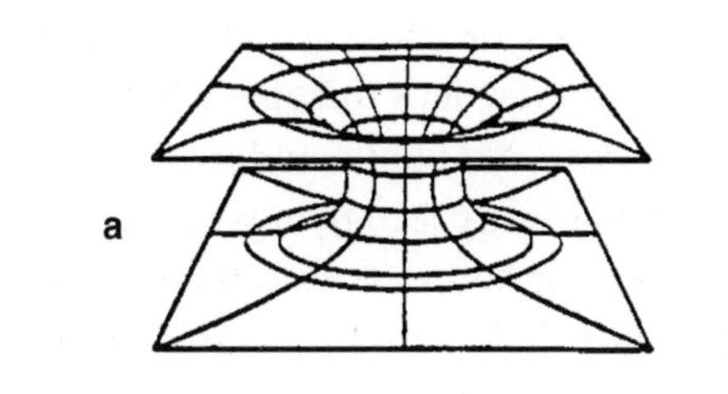

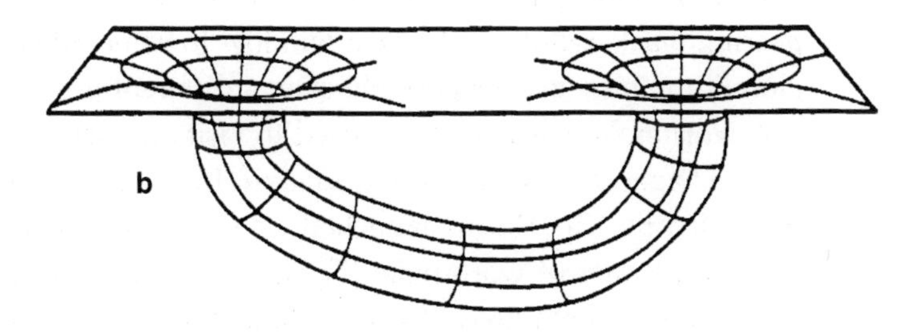

Figure 8.7. Spacetime in the vicinity of wormholes: (a) mouths at same point in space but different times, and (b) mouths at different points in space at the same time. Recall that the time dimension is along the vertical direction.

know if they actually exist. First proposed over 80 years ago, shortly after general relativity was devised, a wormhole is a solution to the field equations in which two distant regions of the universe are connected by a short narrow throat (see fig. 8.7a, b). In 8.7b we see two wormhole mouths at different space points, but on an (x,y) plane for a single common time (time being the vertical dimension); in 8.7a we see two mouths at the same space point, but on different (x,y) planes. It would also be possible for a wormhole to link two points separated in both space and time. Can you imagine what that diagram might look like? If you entered one mouth of the worm hole and survived the trip, you might exit the other mouth at some remote point in space, and at a time in the distant past or future.

 What would a wormhole actually look like if you stumbled upon one? First, let's imagine what a black hole would look like up close. No light could reach you from inside the black hole, so not surprisingly it would appear black. However, since the curvature of spacetime in its vicinity is very great, the black hole acts like a gravitational lens, and it causes a strong visual distortion in its vicinity. The exterior region of a wormhole might also be one of great spacetime distortion. However, the wormhole itself, instead of appearing black, would look like a window to some other region of the universe, or possibly even a different universe.

A wormhole was the device used to visit an alien world in Carl Sagan's book and movie *Contact*. In 1985, a CalTech group led by Kip Thorne launched a serious investigation into the physics of wormholes.[5] They wanted to see just what would be required of a wormhole that would be traversable by a human, meaning that the wormhole should be stable against collapse and also have only modest gravitational fields along the traveler's path.[7] Once these conditions are expressed quantitatively in terms of spacetime curvature, the equations of general relativity tell us what arrangement and type of matter would be necessary to produce that curvature.

What Thorne's group found was that the desired wormhole would need to be threaded with matter or fields of a highly exotic nature. In order to withstand the inward crushing force of gravity, the matter would need to be capable of generating tremendous outward forces. Yet the matter would also need to have extremely low mass or energy den-

[7] The danger for a traveler is not the size of the gravitational forces along his path, but rather their variation along the length of his body, i.e., their tidal effect. Near a black hole, tidal forces are so large that you would be stretched out like a piece of taffy.

 sity, lest it contribute significantly to the inward gravitational attraction contributing to a collapse. Imagine a material having the strength (on a pound-for-pound basis) more than a trillion times that of steel, and you get some idea of the strange nature of this exotic matter.

One way to describe such exotic matter is that it must have negative energy or mass. The idea of negative energy is not unknown in physics. We routinely describe the gravitational potential energy between two masses as being negative. However, here we are speaking of negative total energy (including rest energy). Nevertheless, negative energy has arisen in physics in a variety of contexts.

In 1948, Hendrik Casimir suggested negative energy as a quantum mechanical fluctuation of the vacuum that could be (and has been) observed as an attractive force between a pair of uncharged conducting plates.[6] Negative energy also comes into play in the postulated evaporation of black holes discovered by Stephen Hawking in 1974 (although not yet observed), and also in the theory of cosmic inflation that describes an exponentially rapid expansion of the early universe.[7] Negative energy (and its gravitational repulsion) also may be responsible for the observation that the most distant matter in the universe seems to be moving away from us at an accelerating rate.[8] Finally, faster-than-light particles (tachyons), if they exist, would have negative energy in certain reference frames (see chapter on faster-than-light speeds).

No one knows whether matter exists in a form that would create the negative energy in sufficient concentration needed to stabilize a wormhole, or whether such stable wormholes might actually exist in nature. (Stephen Hawking, for example, has weighed in against the possibility of wormholes being traversable with his "chronology protection conjecture"—so named because it would make the past "safe" for historians.)[9] If Hawking's conjecture is mistaken, however, and stable traversable wormholes do exist, then they are time machines. Surprisingly, even those wormholes that linked to remote points in space at the *same* time could be used for time travel into the past. Let's see how you could dial up any time you wanted by making use of such a wormhole.

The entrance mouth of the wormhole is kept fixed relative to the distant stars, while the exit mouth at some distant location is rapidly moved at very high speed, either back and forth or in a circle. (Wormholes, like any massive objects, can be moved by gravitational forces.) Clocks at the moving mouth advance more slowly than those at the fixed mouth because of time dilation, but if you go *through* the wormhole, the clocks on each mouth show the same proper time. (Effectively, you traverse the wormhole in no time at all.) Figure 8.8 shows how the time machine works.

The fixed mouth of the wormhole moves along the vertical worldline, while the oscillating mouth moves along the zigzag worldline. Notice how successive clock ticks are more widely spaced along the zigzag worldline due to time dilation. (The two clocks were synchronized at time $t = 0$ when the time machine was created.) Consider a traveler who starts out at the fixed mouth (at time $t = 15$ yr), and travels through ordinary space along path A, eventually reaching the moving mouth (where the clock reads t = 6 yr). This part of the traveler's journey is forward in time at sub-light speed (worldline slope no less than 45 degrees). The return trip back to the entrance mouth is made through the wormhole itself (dotted path B), which connects two remote points at the same proper time ($t = 6$ yr). The net result of the round-trip is that you have gone back in time by 9 years and returned to your original location.

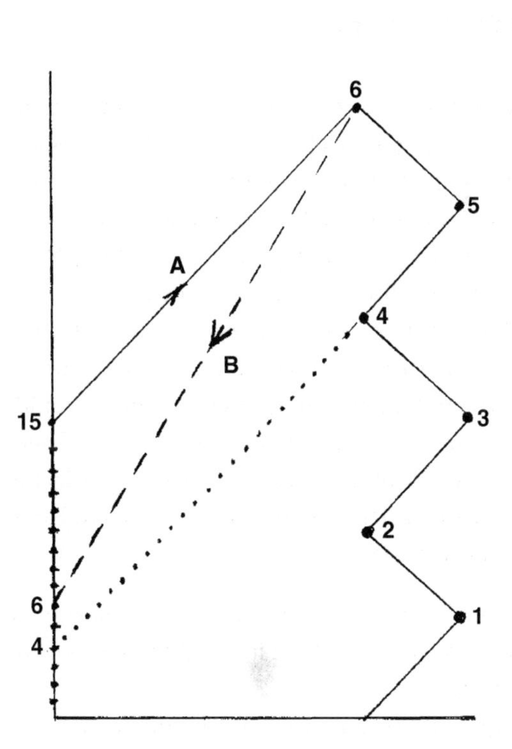

Figure 8.8. Zigzag line shows the worldline of one mouth of a wormhole that oscillates back and forth. The other mouth is stationary and its worldline coincides with the vertical time axis. One-year intervals are shown on each worldline. An example of a backward trip in time is shown: someone leaves the stationary mouth of the wormhole at t = 15 yr, travels through ordinary space along path A at nearly light speed, reaches the other mouth, and returns through the wormhole (path B) at t = 6 yr.

There is an important restriction to the type of backward time travel depicted in figure 8.8. It only allows time travel back to a certain time after the machine has been created. The "time travel boundary" is that time at which a worldline connecting equal proper time points at the fixed and moving mouths makes a 45-degree angle (the dotted line connecting points at t = 4 yr). A traveler starting out at this particular time, who traveled at the speed of light through space to reach the other mouth of the wormhole could then return home through the wormhole, and arrive back at the same time he left.

Why You Couldn't Kill Young Grandpa

Despite the above limitation on how far back future time travelers would be able to travel, all the paradoxes associated with time travel remain. For example, several hundred years after the creation of such a time machine, time travelers could face the prospect of going back and killing their grandfathers. Consider this inherently contradictory sequence: you go back in time → you kill your infant grandfather → you are never born → you couldn't go back in time. Any sort of time travel that permitted such nonsense would seem to be impossible. Science fiction writers have been particularly inventive in finding reasons why it would be impossible for a time traveler to do the dastardly deed. Here are a few of the speculations.

1. Time travel is so dangerous that no one ever survives the trip alive. (Maybe you turn into antimatter as you go back in time.)
2. The first trip back in time destroys the universe (and it hasn't happened yet!).
3. You can only go back to a "random" point in space and time, or a point that is not of your choosing. The chances of your going back to the time and place where your grandfather was is essentially zero.
4. The amount of energy needed to send you back in time more than a few seconds becomes prohibitively great.
5. You are unable to stop at any particular past time, but only quickly sweep past it.
6. After you go back to your grandfather's time you can only observe, not interact with matter.
7. Any changes you bring about by killing your grandpa

have already happened, therefore the person you killed was not really your grandpa after all.

8. You lose all memory of the future when traveling back in time, and have no idea why you are there or where you came from.
9. Every time you try to kill your grandpa something goes wrong: the bullet misses, the gun jams, etc., etc. These events preventing his death must happen because there is only one past, and it cannot be changed.

The last suggested reason why you couldn't go back and kill grandpa is particularly interesting, because it raises the specter of free will being an illusion. But, suppose we consider a version of the grandfather paradox in terms of a round-trip to the *future*. After visiting the future you learn that a descendant of someone alive today will make a decision that destroys the planet. On your return to the present you resolve to kill that person in order to save humanity from his evil descendant. Does it seem just as reasonable to say that you couldn't succeed now as in the grandfather paradox?

On the face of it, you might say no. After all, the future hasn't happened yet, so how can we speak of changing that which hasn't happened? If we have free will, surely we should be able to take an action today that alters a possible future that would happen without our intervention. However, if you believe that, and you also believe that the past is unchangeable (even by a time traveler), you are effectively saying that people alive now have free will, but they didn't have it in grandpa's day. The real issue about not being able to go back and kill grandpa is not any apparent lack of free will, but the question of self-consistency. You couldn't go back and do the deed not because you would lack the will, but because the past didn't happen that way (since you exist), therefore it *can't* happen, no matter how hard you try.

Figure 8.9. THE FAR SIDE © 1987 FARWORKS. All Rights Reserved. Reprinted with permission.

If you are not satisfied with that explanation, here's a tenth possibility: you could go back in time and kill your grandpa, but then you would be in an alternate universe in which he never survived to give birth to your father. So how did you get to go back in time in the first place? You are just a drop-in from some other parallel universe in which grandpa *did* survive. This fantastic idea is consistent with the many-worlds version of quantum theory proposed by Hugh Everett III.[10] In Everett's controversial theory, the universe is continually branching in time, with parallel universes splitting off, corresponding to different possible outcomes to events.

You may find Everett's explanation of the grandfather paradox too bizarre to be credible. The point of the discussion has not been to convince you that there exists a definitive answer to the grandfather paradox. Rather, it was to show that backwards time travel does not automatically lead to the paradox. All we can say for sure is that apart from Everett's many-world's idea, backward time travel *that allowed you to kill grandpa* is impossible. But, since we don't know all the "rules" imposed by the physics of time travel, we cannot say for sure that any of the nine listed reasons are wrong regarding why you (a time traveler) couldn't kill grandpa. If someone claims that time travel is impossible because of the grandfather paradox, the burden of proof must be on that person to show that all of the nine proposed solutions (plus any number of others not listed) are wrong.

My rating for the idea that time travel to the past is theoretically *possible is zero cuckoos. But I give it 2 cuckocs in terms of feasibility in our universe.*

9 Faster-than-Light Particles Exist

OVER a century ago, a young boy dreamed about chasing light beams. Even as a boy, Albert Einstein realized that chasing a light beam would be as futile as chasing a rainbow—an idea that later was at the core of his relativity theory. According to relativity, not only couldn't you catch up to a photon in vacuum,[1] but you couldn't even decrease its speed relative to you by chasing after it. Moreover, it is impossible to accelerate massive bodies to the speed of light in vacuum, *c*, no matter how long they are accelerated. For these reasons Einstein concluded in 1905, when he published his theory of special relativity, that speeds in excess of light "have no possibility of existence."[1] But, despite (or perhaps because of?) Einstein's belief, a few physicists have questioned whether relativity necessarily rules out faster-than-light (FTL) speeds.

One possible way that real particles could have FTL speeds without being accelerated through a light-speed "barrier" was suggested in 1962 by Bilaniuk, Deshpande, and Sudarshan.[2] They reasoned that a hypothetical particle, later called a *tachyon*,[2] might be FTL from the very moment of its creation in some subatomic particle reaction, and that it could never have speeds below *c*. By having particles "born" with FTL speed, we avoid the problem that it would take an infinite amount of energy to accelerate a particle up

[1] A photon can be thought of as a particle of light, or a packet of light energy.

[2] Tachyon is from the Greek *takhus*, meaning swift.

Figure 9.1. Einstein preparing to chase a light beam. Permission to use illustration courtesy of the artist: Jim Warren, www.jimwarren.com.

to the speed of light. The speed of light could then be a two-way speed barrier to particles on either side of it. One advantage of this scheme is that tachyons would fill an otherwise empty class: particles that always move faster than light, and hence they would complement the other two known classes of (1) particles that always travel at light speed, such as photons and gravitons, known as *luxons*, and (2) those that always travel at sublight speed, known as *tardyons*. We can see an interesting symmetry between the three classes of particles (tachyons, tardyons, and luxons) by plotting the momentum versus energy for each type of particle (see fig. 9.2).

Tachyons, *if they exist*, would not be the first particle ever suggested on theoretical grounds before they were observed. In fact, a long history of such theoretical conjectures that preceded experimental findings include the positron, antiproton, pion, Omega Minus, Z and W bosons, three types of neutrinos, and six types of quarks. Some readers

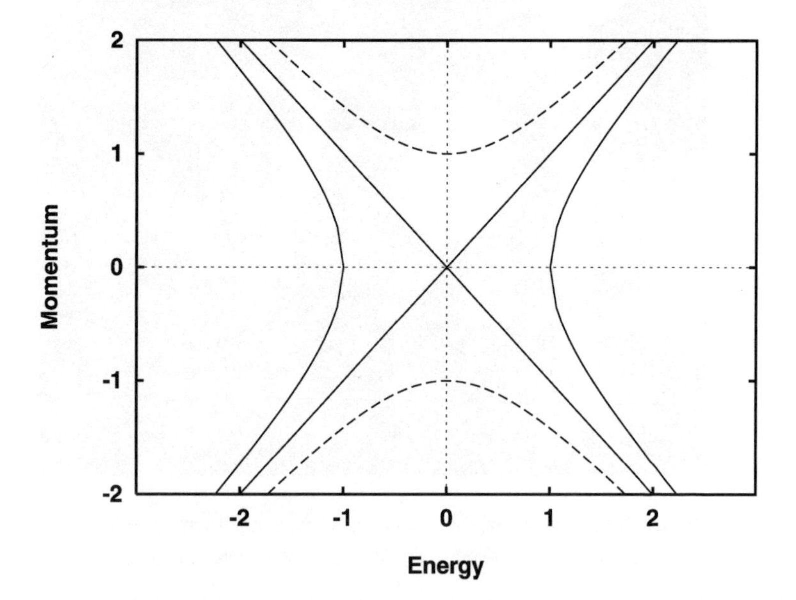

Figure 9.2. Momentum (p) versus energy (E) for tachyons, tardyons, and luxons. E is in multiples of the rest energy, mc^2, and p is in multiples of mc. p and E are related so as to fall on the two straight lines for luxons, the dashed hyperbola for tachyons, and the solid hyperbola for tardyons. As the figure shows, tardyons have a minimum possible energy $E = 1$, but no minimum momentum, while tachyons have a minimum possible momentum $p = 1$, but no minimum energy.

175 may object to such a comparison because the listed particles were either part of some theoretical model or, in the case of the neutrino, were postulated based on empirical evidence. In addition, tachyons are not just some other type of particle; in fact, as a result of their FTL speed, tachyons have a host of bizarre properties. These properties all follow from extending the equations of relativity to the realm of "superluminal" ($v > c$) velocities. In order to keep the discussion of these tachyon properties at the conceptual level, we shall place sketched-out derivations associated with each property in footnotes rather than in the text itself.

Properties of Tachyons

1. Imaginary rest mass. Imaginary numbers by definition are the square root of negative numbers. Saying that the rest mass of a tachyon, m, is imaginary means that the quantity m^2 is negative. All observable physical quantities, such as energy and momentum, must have real rather than imaginary values. But, in order that tachyons have real values for their energy, E, and momentum, p, the equations of relativity require that the tachyon's rest mass must be imaginary and hence unobservable.[3] An imaginary unobservable rest mass is not a problem, however. Just as with photons (packets of light), the rest mass of tachyons is not directly observable because you (being made of tardyons) could never catch up to a tachyon and see it at rest.

2. Futility of chasing tachyons. Let's say you attempted to catch up to a tachyon moving at some fixed FTL speed. If you started chasing after it by continually increasing your

[3] Given $E = \gamma mc^2$ and $p = \gamma mv$, if $v > c$ then $\gamma = 1/\sqrt{1 - v^2/c^2}$ will be imaginary, and hence m must also be imaginary in order for E and p to be real.

 speed, the futility of your chase would soon become readily apparent. The more you increased your speed in chasing the tachyon, the *greater* its speed would be relative to you. This is just the opposite of a "normal" particle, i.e., a tardyon, which would appear to slow down the faster you chase after it.[4] It would be almost as if a tachyon could sense your pursuit and speeded up to elude you. But the tachyon speedup is not the result of any tachyon "consciousness"; it is merely an artifact of the strange nature of how velocities are added or combined when one of the velocities is greater than the speed of light. Not only does a tachyon appear to speed up the faster you chase it, but at a certain chase speed the tachyon would be moving away from you at infinite speed—it would be literally everywhere at once.[5] (Trying to catch up to a tachyon is just as futile as trying to catch up to a photon, which always has the same speed, no matter how fast you chase it.)[6]

3. Energy decreases with increasing velocity. The relationship between the total energy of a particle, E, and its velocity, v, is shown in figure 9.3 for the entire range of velocities—both $v < c$ (tardyons) and $v > c$ (tachyons).[7] Given

[4] If you chase after a tachyon of speed v with a chasing speed of u, then the tachyon's speed relative to you will be $v' = (v - u)/(1 - uv/c^2)$. $v' > v$ when $v > c$.

[5] v' becomes infinite using the formula in footnote 4 when $u = c^2/v$. If you continued to accelerate after it, the tachyon speed would be finite—but now in the opposite direction. Your further acceleration in the same direction as before (now heading *away* from the tachyon) would make the tachyon appear to slow down relative to you, the faster you moved away from it, but its speed would always exceed c.

[6] Photons travel at the speed c no matter at what speed you chase after them. To see this, substitute $v = c$ in the formula in footnote 4, and you'll find that $v' = c$ regardless of the value of u.

[7] Given the formula in footnote 3, E increases as v decreases if $v > c$. In the limit of infinite v, E will approach zero.

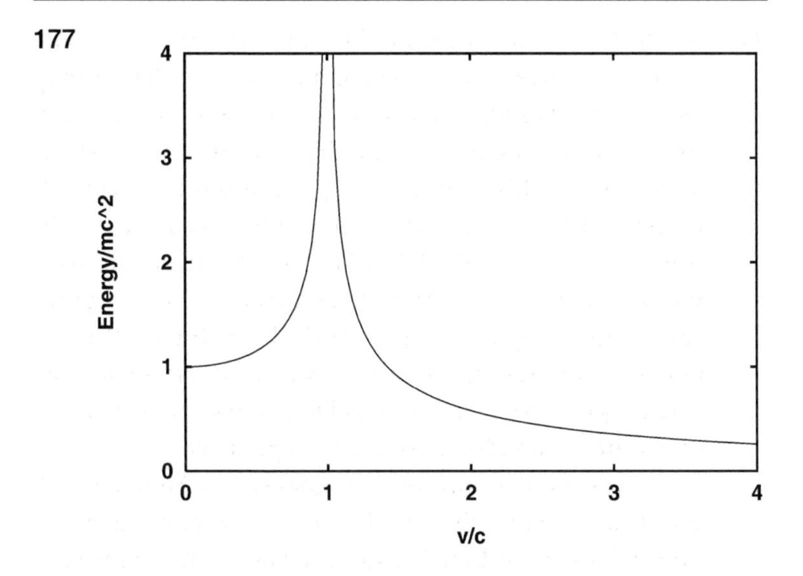

Figure 9.3. Total energy *E* versus *v/c* in units of $|mc^2|$.

the shape of the graph for $v > c$ in figure 9.3, as a tachyon loses energy (as E decreases), it would have to speed up rather than slow down. As the tachyon energy approached zero, its speed would approach infinity and the tachyon would approach the so-called transcendent state. One result of the strange dependence of energy on velocity is that force and acceleration vectors go in opposite directions for tachyons—if you want to make a tachyon speed up, just try to hold it back![8]

4. Negative energy. Suppose you on Earth and an alien in a spaceship both observe a tardyon (perhaps another spaceship) and measure its energy. The two of you will usu-

[8] However, you shouldn't conclude from this fact that the tachyon behaves like a negative mass particle, because $F = ma$ is only applicable in the case of an object having fixed mass. As the tachyon slows down, its relativistic mass increases, so the net result of a force is to increase the tachyon momentum in the direction of the force.

 ally find different values for the tardyon's energy, but both of you will always conclude that the tardyon's *total* energy (which includes its kinetic plus rest energy) must be positive. But now, suppose you and the alien are both observing a tachyon. In this case, you might measure the tachyon's energy to be positive, and the alien might measure it to be negative—or vice versa.[9] The idea that tachyons can have positive energy according to some observers, and negative energy according to others, is particularly strange. Because of this weird property, certain reactions that violate the law of energy conservation can actually take place in some reference frames—an idea explored at length later.

5. Back to the future? The connection between FTL speeds and reverse time has become a part of popular culture and science fiction; it also has a basis in relativity. Suppose you on Earth use a transmitter to send a message using tachyons having speed $v > c$ to a friend in another galaxy.[10] Aliens in a spaceship heading toward the other galaxy might find that the time order between the transmission and reception of the message gets switched—the message would be literally received before you sent it![11]

Under what conditions would the aliens observe this bizarre result? Their speed away from Earth would have to

[9] If one observer sees a tachyon of energy E and velocity v, then a second observer with relative velocity u in the direction of the tachyon's motion will measure its energy to be $E' = \gamma(E - up)$. E and E' will have opposite signs if $u > E/p = c^2/v$.

[10] Any device that emitted tachyons could serve as a message transmitter. By simply turning the device on and off, you could create a stream of bits (zeroes and ones) that comprised a message.

[11] A tachyon transmitted from Earth at time $t = 0$ will be received at a galaxy x light years away at a time $t = x/v$. According to aliens in a spaceship heading toward the galaxy at a speed u, the signal will be received at the galaxy at a time $t' = \gamma(t - ux/c^2)$, which is less than zero for $u > c^2/v$.

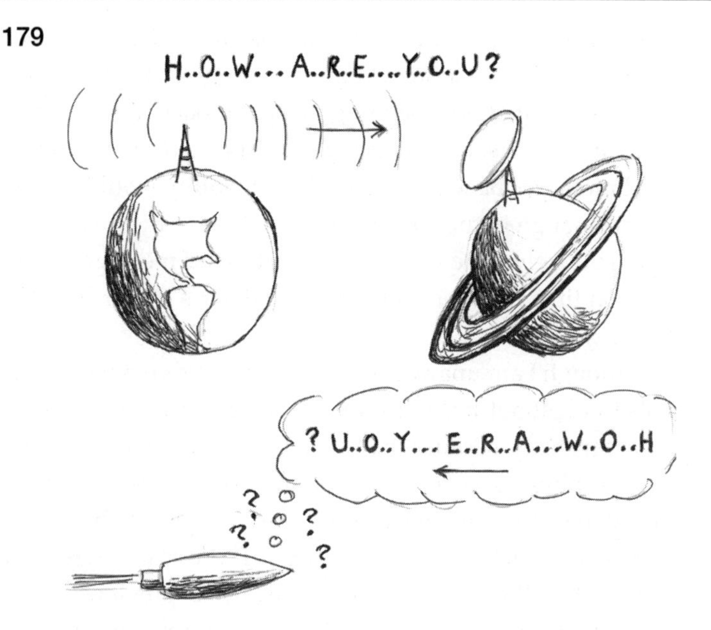

Figure 9.4. Tachyon message according to two different observers. Earth observers regard themselves as the transmitters of the message, while observers in the spaceship regard the Earthlings to be the recipients of the message.

be greater than $u = c^2/v$, which is less than c. (For example, if your transmitter on Earth were using tachyons whose speed was ten times the speed of light, the aliens in the ship would see time reversals occurring when their ship's speed was at least a tenth the speed of light.) Note that both the reversal in time order and the reversal in the sign of the tachyon energy, discussed earlier under property 4, occur at the same minimum speed of the alien ship.

Physicists refer to reversals in the time order of a cause (message transmission) and its effect (message reception) as violations of causality. One possible way to accommodate such reversals without paradox is to reinterpret who was the sender and the receiver of the message, or which was the cause and the effect. Opinion is divided on whether such a reinterpretation scheme can truly avoid the paradoxes associated with sending messages back in time—say, to your earlier self.[3] If tachyons really did allow you to send messages back in time, such a capability would probably be enough in most people's eyes to make their existence *highly* unlikely. (The use of tachyons to send messages backward in time is a theme of John Carpenter's horror film *The Prince of Darkness*.)

Although I am unaware of any poll results on what physicists think about tachyons, probably a majority do not take them seriously, because they are so weird (i.e., the tachyons, not the physicists). This skepticism has been further reinforced by the negative results of experiments that have searched for tachyons over the years. These searches rely on one or the other tachyon properties listed previously. For example, some searches, in which a particle's speed is directly measured, have looked for FTL speeds of cosmic rays or accelerator-produced particles. Other experiments have looked for particles having imaginary mass or negative m^2 (which can be found from measured values of the particle's energy and momentum).[12] Apart from one unconfirmed observation, based on a single cosmic ray event, no tachyons have been seen.[4] Of course, it is always possible that tachyons exist but that they do not interact with other matter. However, such a possibility would put tachyons in the metaphysical realm.

In view of all the negative experimental evidence, why do some physicists still think that tachyons might exist? First, because experiments cannot prove that a particle doesn't exist, only that, if it does exist, its properties are such that

[12] The mass of a particle can be found from its momentum, p, and energy, E, using the formula $m^2 = E^2/c^4 - p^2/c^2$.

 it would have escaped detection under particular experimental conditions. In addition, among the known particles, there is one category, the neutrinos, which have masses sufficiently close to zero so that the value of m^2 for all three types of neutrinos could well be negative, given the measurement uncertainties.

Could Neutrinos Be Tachyons?

Originally, only one type of neutrino was known—that associated with the electron, known as the electron neutrino. Now, we know that neutrinos come in two other types that are associated with the muon and tau particles. The neutrino was first postulated by Wolfgang Pauli in the 1930s to explain certain puzzling features of nuclear beta decay. Pauli was a brilliant theoretical physicist who had a reputation for making equipment go haywire in his presence. Experimental physicists were said to fear letting Pauli into their labs because of this "Pauli effect." Fortunately, Pauli's ability to analyze the results of experiments was far ahead of his facility with equipment. Let's see how he realized that an *unobserved* particle (the neutrino) had to be present in beta decay. Suppose a "parent" nucleus A were to decay or disintegrate into a "daughter" nucleus B and an electron (the "beta" particle). We could represent this decay process as $A \rightarrow B + e$. Since the heavy daughter nucleus B recoils with very little energy, the light electron would get almost all the energy E released in the decay, where $E = \Delta mc^2$. (Δm is the mass difference between the parent and daughter nuclei.) In other words, if nuclei decayed only into two particles, we would find that the electron is always emitted with a specific well-defined energy, and the result would be a single *line* spectrum.

 The observed spectrum in beta decay, however, is not a single line, but rather a *continuous* distribution of electron energies. Pauli realized that such a continuous distribution required that a third unobserved particle (the neutrino) had to be present in beta decay. We could then write the decay as $A \to B + e + \nu$. Pauli believed that a third particle was present, because with a third particle sharing the released energy, the electron's energy could vary from one decay to the next, and a continuous spectrum was the result. The name neutrino (Italian for "little neutral one") was suggested by Enrico Fermi, because the particle seemed to have a very low mass, and it had no electric charge. Even though the neutrino was not directly observed in the experiments of the 1930s, its mass could be found indirectly by observing what was the highest energy electron emitted in beta decay, or by finding the endpoint of the energy spectrum. (The idea is that the smaller the neutrino mass, the greater the energy left over for the electron, and hence, the higher the endpoint energy.)

The simplest form of beta decay is that of a free neutron, which decays according to $n \to p + e^- + \bar{\nu}_e$ where $\bar{\nu}_e$ is the antineutrino, or the antiparticle of the neutrino, and the subscript e identifies it as the neutrino associated with the electron. The most precise estimate for the mass of the electron neutrino (or antineutrino) comes from the beta decay spectrum for tritium, which is an isotope of hydrogen. These experiments actually measure the square of the mass (m^2) rather than the neutrino mass itself. Over the years, most tritium beta decay experiments have given negative values for m^2, in some cases by over five standard deviations.[5] If statistical uncertainties were the only source of measurement error, such results would be fairly conclusive evidence that the electron neutrino was a tachyon, since only tachyons have an imaginary rest mass. However, experimenters, being well aware of possible sources of systematic error in these difficult experiments, have *not* claimed that these negative m^2 values show that the electron neutrino is a tachyon.

If electron neutrinos really are tachyons, what other consequences would follow, aside from a small modification of the endpoint of the beta decay spectrum? In 1992, Chodos, Kostelecky, Potting, and Gates suggested a most dramatic test of the tachyonic neutrino hypothesis involving the decay of stable particles such as the proton.[6] Consider the "decay" $p \rightarrow n + e^+ + \nu_e$ where e^+ represents a positron, or antielectron. Such a process can occur for a proton inside some nuclei, but it is energetically forbidden for a free proton, given that the neutron is more massive than the proton. However, if the neutrino is a tachyon, recall that its energy can be negative in certain reference frames and positive in others (see property 4 of tachyons). This means that you (in the lab reference frame) could observe a positive energy neutrino being emitted by a high energy decaying proton, even though someone moving with the proton would say the neutrino energy is negative.

How would someone moving with the proton (for whom it is at rest) interpret an emitted negative energy neutrino? This observer sees an antineutrino *absorption* (by the proton) rather than a neutrino emission—a reversal in time. In other words, if you were riding with the proton, you would think the process is not a proton decay, but instead the reaction $\bar{\nu}_e + p \rightarrow n + e^+$.[13] This switch from an emitted (outgoing) to an absorbed (incoming), particle depending on the observer's reference frame—a violation of causality—is based on property 5 of tachyons discussed earlier (see fig. 9.5).

[13] The absorbed antineutrino is presumably from some background sea of particles that fills all space.

The reinterpretation of negative energy backward-in-time antiparticles as positive energy forward-in-time particles avoids a number of theoretical difficulties with tachyons. For example, if tachyons can have negative energy, you

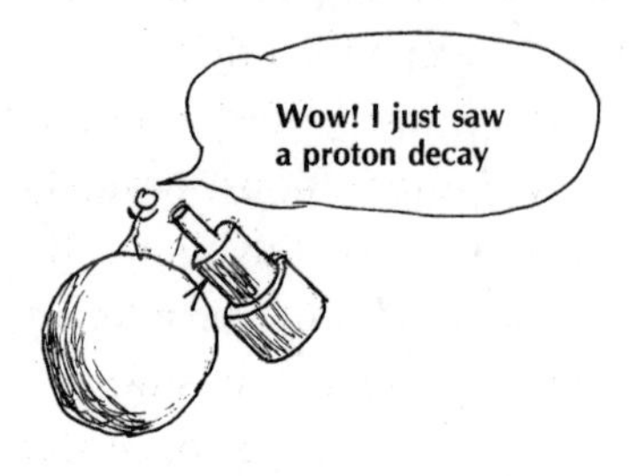

Figure 9.5. An incoming antineutrino becomes an outgoing neutrino, depending on your reference frame.

might think that would imply the vacuum is unstable. This is because a tachyon-antitachyon pair could spontaneously materialize from nothing, and still conserve energy and momentum, as long as the two particles had equal and opposite energy and momenta. However, if the negative energy antitachyon member of the pair is reinterpreted as a positive energy tachyon traveling in the opposite direction, we see that the point where the pair materialized from nothing in figure 9.6 is actually just an arbitrary point on the path of a single tachyon.

Suppose it is true that neutrinos are tachyons, which requires that protons can decay once they have a sufficiently high energy. Unfortunately, it is not possible to predict what the minimum energy is for proton decay without knowing the mass of the neutrino. In fact, the predicted threshold en-

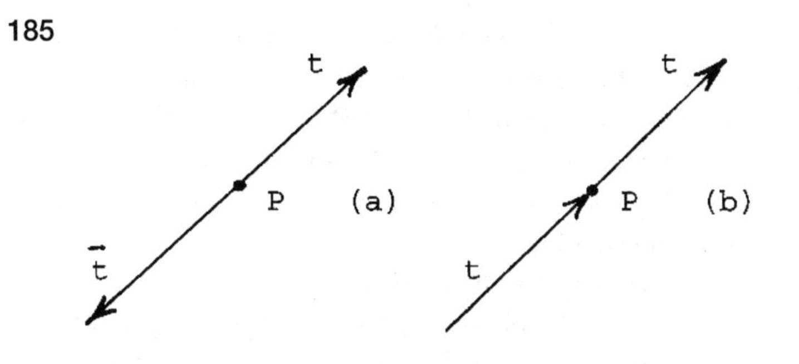

Figure 9.6. In (a) we see the "decay" of a vacuum into a back-to-back tachyon-antitachyon pair occurring at point P. But the negative energy (reverse time) antitachyon $\bar{t}$ can be reinterpreted as a positive energy tachyon t traveling forward in time. As we see in (b), point P is in reality just an arbitrary point on the path of a single tachyon.

ergy for proton decay depends inversely on the absolute value of the mass of the electron neutrino.[7] Thus, one must look at very high energy protons in order to test the tachyonic neutrino hypothesis, given the minuteness of the neutrino mass. No proton decays, of course, have been observed up to the highest energies attained in particle accelerators. However, cosmic ray particles (which include mostly protons) occasionally bombard the Earth with energies up to a million times higher than those in the highest energy particle accelerators. So, cosmic rays would seem to offer the most sensitive test of the hypothesis that the electron neutrino is a tachyon.

Might Proton Decay Be Seen in the Cosmic Rays?

Given the rarity of very high energy cosmic ray protons, the idea is not to look for decays of individual particles, but rather to see what effect proton decay might have on the shape of the cosmic ray spectrum, i.e., the numbers of cosmic rays having various energies. In particular, we want to see if there is any evidence for a sudden depletion of protons from the spectrum above some particular energy—the threshold for proton decay. Unfortunately, little is known about the mechanism for creating the highest energy cosmic rays and what their original spectrum is, so the model is highly speculative.

The observed spectrum of cosmic rays above an energy $E = 10$ GeV (10 billion electronvolts) roughly follows a power law $1/E^N$, where N is approximately equal to 3. This power law means that for every factor of ten increase in cosmic ray energy, the number of particles per unit energy interval decreases by a factor of $10^3 = 1000$. Obviously, as a result of this power law, the higher in energy you look, the larger the area detectors must cover to observe a significant number of increasingly rare cosmic ray events. At the highest energies at which cosmic rays have been observed (over 10^{20} eV), they are so rare that even with the use of detectors covering many square kilometers, only a few events are seen per year. At these highest observed energies, individual protons have as much energy as a pitched baseball.[14]

Suppose we examine the cosmic ray spectrum on a log-log plot (see fig. 9.7). In such a plot a power law like $1/E^N$ appears as a straight line with slope –N. One curious feature of the observed cosmic ray spectrum in the figure is the "knee" (steepening of the slope) that occurs at an energy of about 4.5 PeV, or log $E = 15.6$. (A PeV or a Peta eV is 10^{15} eV.) In most conventional models for the production of cosmic rays, this fairly abrupt change in power law at the knee is assumed to be due to the transition from one type

[14] We normally don't think of a pitched baseball as having a huge amount of energy, but remember that a baseball has perhaps 10^{26} protons, and we are saying that a *single* cosmic ray proton has the same energy as all of the 10^{26} protons in a baseball.

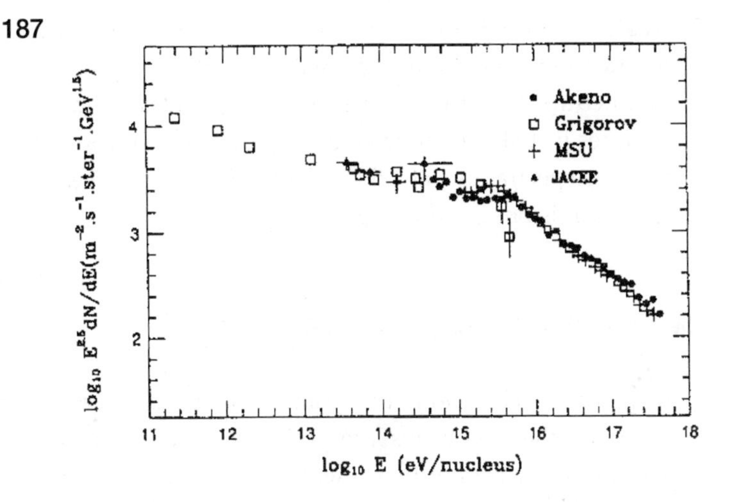

Figure 9.7. The observed cosmic ray spectrum, or the rate of cosmic ray particles per unit area versus energy. Note the use of log scales for both variables. (Don't worry about the strange units used on the *y*-axis.) The "knee" of the spectrum occurs at $E = 4.5$ PeV $= 4.5 \times 10^{15}$ eV, or $\log E = 15.6$. Data are from four groups identified, shown in Gaisser, 1995.

of cosmic ray source to another. But, if instead we assume that cosmic rays have a single power law at their source over the entire energy range above 10 GeV, then the knee could be interpreted as the onset of a depletion of protons from the spectrum due to proton decay. According to the hypothesis that the electron neutrino is a tachyon, protons should be able to decay at 4.5 PeV if the neutrino mass were around $|m| = 0.5 \text{ eV}/c^2$, where the absolute value $|m|$ is defined as $m=\sqrt{-m^2}$.

This highly speculative interpretation of the knee of the cosmic ray spectrum was first raised by Alan Kostelecky in 1992.[8] More recently, I have attempted to model the cosmic ray spectrum over its entire energy range above the knee, given the hypothesis that the electron neutrino is a tachyon.[9] Given the speculative nature of the model, a good fit to the observed cosmic ray spectrum offers only weak evidence that the neutrino is a tachyon. Nevertheless, the fit to the spectrum does make one strange prediction: that neutrons exist in the cosmic rays in a narrow range of energies just above the knee of the spectrum—essentially a neutron spectral line at 4.5 PeV or wherever the knee of the spectrum is observed in a given experiment.

Is There a 4.5 PeV Neutron Line in the Cosmic Rays?

If such a line is found in the cosmic rays, it would be strong evidence that neutrinos are tachyons, because neutrons should otherwise not be present in the cosmic rays at this energy. Given the relativistic time dilation effect, free neutrons (which have a lifetime of around 600 sec) can reach Earth from greater distances, the higher their energy. But, even at an energy as high as 4.5 PeV, neutrons should not be able to reach Earth from much farther than 100 light years before decaying. Such a distance—about a thousandth of the diameter of our galaxy—is believed to be far less than that of most cosmic ray sources. Under the tachyonic neutrino hypothesis, however, neutrons can survive their journey from much greater distances than 100 light years, because they are continually being created from protons en route to Earth via a chain of decays: $n \to p \to n \to p \cdots$ (see fig. 9.8). The result of this decay chain is a "pile up" of neutrons at a particular energy just above the threshold for proton decay—hence the predicted 4.5 PeV neutron line. (Even though the cosmic rays are predicted to spend part of their time en route to Earth as protons and neutrons during their decay chain, the model assumes that the time spent as neutrons is much greater than the time spent as protons.) Thus, in summary, an observation of neutrons among the cosmic rays—particularly a neutron line at a single energy—would be a highly unexpected finding (within the framework of conventional theory), and it would offer strong support for the tachyonic neutrino hypothesis.

The evidence as to whether a neutron line actually exists at 4.5 PeV in the cosmic rays is tantalizingly suggestive at the time of this writing.[10] Before we look at the evidence, let's see how cosmic ray neutrons might be detected. Any particle detector, such as a Geiger counter, can detect the passage of charged particles. In fact, about half the random clicks you would hear from a Geiger counter are due to the passage through the instrument of charged particles created by the cosmic rays—the rest being due to other "background" radiation. At sea level, however, few of the particles that a Geiger counter records are the "primary" cosmic rays themselves. Instead, primary cosmic ray particles, which collide with atoms of the Earth's atmosphere, create a large number of secondary particles that comprise an "air shower." These air showers, which can include billions of particles, descend in the form of a narrow cone whose axis

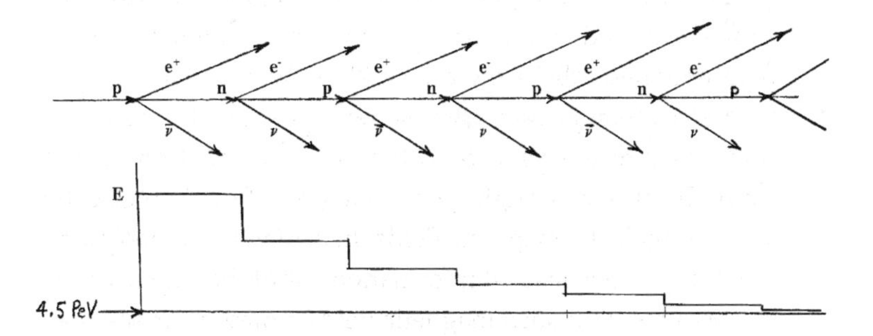

Figure 9.8. Illustration of the $n \rightarrow p \rightarrow n \rightarrow p \cdots$ decay chain that is predicted to occur for protons above the threshold energy of 4.5 PeV. The graph below the decay chain shows how the energy of the neutron or proton is steadily reduced as a result of each decay. The chain stops at 4.5 PeV, leading to a neutron line in the cosmic rays at this energy. Finding such a neutron line at 4.5 PeV would be convincing evidence that the neutrino is a tachyon.

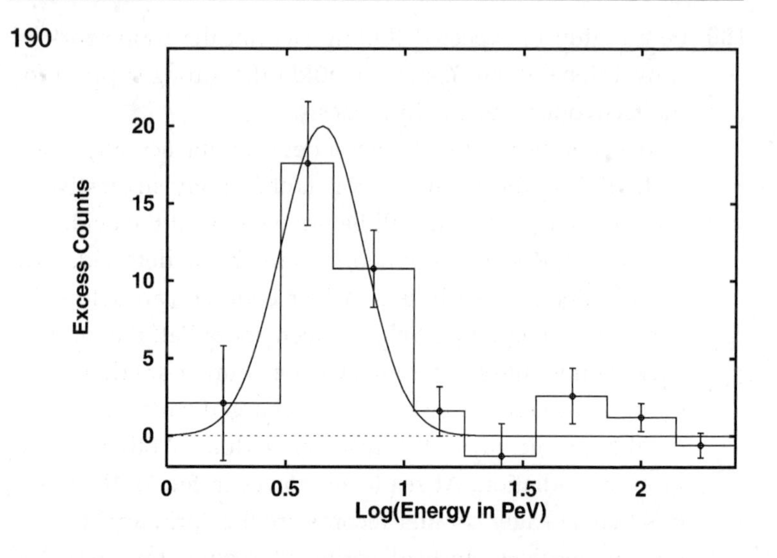

Figure 9.9. Energy spectrum of the excess counts that point back to Cygnus X-3, as reported in the reference in endnote [10].

points in the direction of the sky from which the primary cosmic ray particles arrive.

Surprisingly, it is not easy to distinguish between air showers caused by cosmic ray neutrons and protons at very high energies. About the only distinguishing characteristic between them is that only showers caused by neutrons point back to their source in the sky, since neutrons travel through space in straight lines, being unaffected by galactic and extragalactic magnetic fields. In contrast, charged particles like protons lose all information about the source direction as a result of such magnetic fields, unless their energies are much higher than 4.5 PeV. Thus, any concentration of cosmic rays pointing back to particular directions in the sky would be strong evidence of neutral particles from a source in that direction.

Currently, the conventional wisdom is that there are no sources of neutral particles at very high energies in the cos-

191 mic rays. However, in the 1970s and 1980s there was a flurry of experiments claiming to see such signals from two X-ray binary stars, Hercules X-1 and Cygnus X-3.[11] (Recall that binary stars are pairs of stars in orbit about one another.) Usually, a signal was seen only when selecting data contained in a narrow window of the "orbital phase," i.e., a certain point in the orbit of the binary star. The cosmic ray events pointing back to the two X-ray binary sources appeared to be strongly interacting particles, such as neutrons rather than gamma ray photons, based on their association with copious numbers of muons. In all but one of these experiments, the data were sufficiently sparse that they could not be used to test whether there was evidence of a neutron line at 4.5 PeV.

However, in the one experiment from this era by Lloyd-Evans et al.[12], which did have enough data to look at an energy spectrum, a line does in fact appear at precisely the energy predicted (see fig. 9.9)![10] What makes the entire situation rather murky, however, is that subsequent experiments conducted with much greater sensitivity have failed to see such signals pointing back to Hercules X-1 or Cygnus X-3.[13] This suggests one of two disquieting possibilities: either (1) those sources turned off just about the time that improved instrumentation became available; or (2) all the prior experiments that observed signals were in error.

Elsewhere we have suggested that the first of these two possibilities is probably the correct one.[10] We have also suggested a way to test the hypothesis that there are cosmic rays pointing back to particular sources without waiting for any specific sources to turn back on. The idea is to select only those cosmic ray events in a narrow energy interval centered on 4.5 PeV—or wherever the spectrum knee appears in a given experiment—and look at their arrival directions on a two-dimensional map of the sky.

Figure 9.10. Photo of the author with his "ban tardycentrism" T-shirt.

The prediction of our tachyonic neutrino model is that such events caused by neutrons should strongly cluster about particular points in the sky (the locations of the sources), whereas events in other energy intervals should show no such clustering. The data necessary to conduct this test currently exist, but at the time of this writing the test had not yet been made. Most likely, the data will yield a negative result, and one more revolutionary but highly speculative idea will bite the dust. On the other hand, a positive result would be strong evidence that the electron neutrino is a tachyon, and the world of conventional physics may be thrown into some considerable turmoil. For further developments on the issue, see the Web site http://physics.gmu.edu/~e-physics/bob/tachyons.htm. Regardless of the outcome, physicists need to keep an open mind about the possible existence of faster-than-light tachyons. Since their existence is clearly an experimental question, we should not let any prejudice against the idea, i.e., any "tardycentrism," make us blind to possible ways that tachyons might reveal themselves.[15]

My rating for the idea that faster-than-light tachyons actually exist is zero cuckoos.

[15] Some researchers measuring the square of the electron neutrino mass (which in some experiments has been reported as being negative by five standard deviations) may be guilty of "tardycentrism" by dismissing the idea that the neutrino is a tachyon. For a discussion on this point, see R. Ehrlich, http://xxx.lanl.gov/abs/hep-ph 0009040.

10 There Was No Big Bang

DID OUR universe have a beginning, or was it always here in much the form we see it today? These competing alternative ideas have existed throughout most of human intellectual thought, and they are intimately tied to our ideas of the existence of a Supreme Being. It has only been in recent times, however, that humans have been able to rely on scientific data rather than philosophical speculations in order to answer questions about the origin of the universe. The idea that the universe began with a big bang around 10 to 15 billion years ago and has been expanding ever since is now generally accepted by nearly all mainstream cosmologists, and by much of the general public. The term "big bang" actually started out as a term of ridicule coined by skeptics of the theory. How did the idea of a big bang go from a derisive term to a generally accepted theory, and why do a handful of cosmologists still believe that the big bang never happened?[1]

Hubble's Law

We can understand the evidence for the big bang by imagining what happens during an explosion in space. The parti-

[1] We have not considered here the challenge to the big bang theory posed by the creationist idea that the world was created about 5,000 years ago, because as noted in the introduction, this idea is "unfalsifiable" and therefore outside the domain of science.

cles of matter blown apart in an explosion would move outward in straight lines at various speeds, with the faster particles traveling greater distances at any given time. In fact, at any given time T, each particle of matter will have traveled a distance r that is proportional to its speed v, where $v = Hr$, with H a constant of proportionality. On the scale of the whole universe, we can think of the galaxies, which typically have 100 billion stars each, as being simply particles of matter. If the galaxies were actually expelled in a big bang, their distances and speeds relative to their point of origin would also obey the equation $v = Hr$ just described.

This relation $v = Hr$ is essentially what Edwin Hubble found when he made measurements of distant galaxies (see fig. 10.1). In fact, the equation $v = Hr$ is known as Hubble's law, and H is known as Hubble's constant. The time T since the big bang is the reciprocal of H, assuming constant galactic speeds during the expansion.

The preceding description, of course, greatly simplifies Hubble's discovery in a number of respects. First, we cannot think of the big bang as an explosion into a preexisting space away from some identifiable point, but rather space and time were *created* in the big bang. Thus, we cannot point to any one spot in space and say that is where the big bang occurred. The galaxies are all flying away from Earth, but that doesn't make Earth the "center" of the explosion. An alien in any galaxy would see all the others moving away from it just as we do. The situation is similar to the way in which dots on a balloon would all move away from each other as the balloon inflates. Although, unlike the dots on an inflating balloon, galaxies themselves do not become larger as the universe expands, and of course galaxies did not exist at all until billions of years after the big bang. A second simplification is that the linear relation between dis-

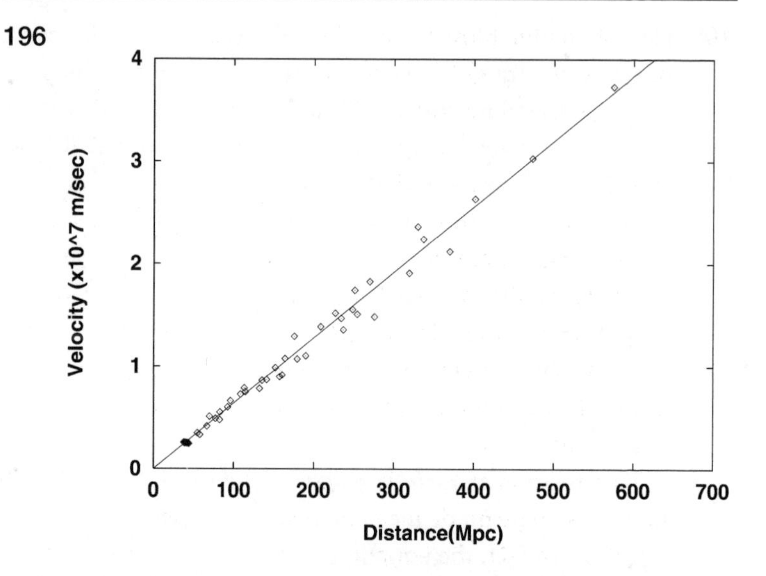

Figure 10.1. Hubble's law showing the linear relation between recession velocities and distances of galaxies. The straight-line fit to the data gives a Hubble constant of 64 km/s per megaparsec (Mpc), or in more conventional units, H = 1/T, with T = 15 billion years. The data are from Filippenko and Riess, 1998.

tance and speed is not an exact one, because galaxies also have a local component of motion in addition to taking part in the overall expansion. As a result, some nearby galaxies are actually moving toward Earth rather than away from it. (This local motion is what causes the data in figure 10.1 to have some jitter about the straight line fit.)

A third complication is that galactic distances are so great that even though they may be moving quite rapidly, we never actually observe any galaxies change their location within a human lifetime. Their motion must be inferred by indirect means. Galactic speeds toward or away from Earth are measured based on the shifts in frequency of light from them due to their motion relative to us. These Doppler shifts are referred to as *redshifts* for galaxies receding from Earth and *blueshifts* for galaxies approaching Earth.

 Galactic distances are even more difficult to measure than their speeds. Essentially, we need to observe an object of known brightness, sometimes called a "standard candle," contained within the galaxy and infer the galaxy's distance based on how bright the object appears to us. We then use the well-known inverse square law to obtain the galaxy's distance. Many galactic distances are so great that only the brightest standard candles can be observed, such as when a massive star undergoes a supernova outburst.[1] Here's an example of the method for reckoning relative distances. Suppose a supernova in galaxy A appears to us to be one hundred times as bright as one in galaxy B. Then, by the inverse square law, galaxy B is ten times farther away than galaxy A—assuming that distance is the only factor affecting the apparent brightness.

Another simplification in our original presentation of Hubble's law was the assumption that the galaxies flew apart with an unchanging speed. If instead the galaxies have been slowing down during the universal expansion due to their mutual gravitational attraction, the age of the universe T would be less than $1/H$. That's very easy to understand: just imagine a runner who runs a one-mile race and who slows down before the finish line. If we only observed his reduced speed at the finish line and assumed that he ran the entire race at that speed, we would compute a longer time for the mile run than his actual time.

Cosmic Background Radiation

Aside from Hubble's law, what other evidence is there in support of the big bang theory? A second line of evidence is the microwave radiation that fills all space. This cosmic background radiation has been poetically called the last faint whisper of the big bang. According to big bang theory, electromagnetic radiation (light) and matter were completely interspersed initially in a primordial fireball of incredibly high temperature. Even after the fireball expanded and cooled for 100,000 years to a temperature of about 3000°K, and hydrogen atoms could form without disintegrating in collisions, radiation still couldn't travel very far before it was absorbed by the atoms. But, once the universe expanded and cooled below 3000°K, light was able to travel vastly greater distances before being absorbed by matter. The universe had become transparent. The cosmic background radiation has been expanding and cooling ever since that time, and by now its temperature should be around 3°K, or three degrees above absolute zero, according to the big bang theory.

The empirical observation of the cosmic background radiation was a great triumph for the big bang theory, even though the discovery was made by accident. In 1964, Arno Penzias and Robert Wilson, a pair of scientists at Bell Labs, were studying ways to improve microwave antennas for communications purposes—work completely unrelated to astrophysics. Despite their best efforts, they were unable to rid their equipment of a constant noise signal that seemed to come from all directions. At one point, they even thought that pigeon droppings on the antennas might be to blame for the noise.

Penzias and Wilson had stumbled upon the cosmic background radiation (a discovery for which they deservedly received the Nobel Prize in 1978). Measurements made on this background radiation over the years have shown that its spectrum perfectly fits that of a blackbody, which applies to objects that absorb all wavelengths equally (see fig. 10.2). The temperature for a blackbody spectrum can be easily found in terms of the wavelength of the peak of the spectrum—raise the temperature of a body, and the peak shifts

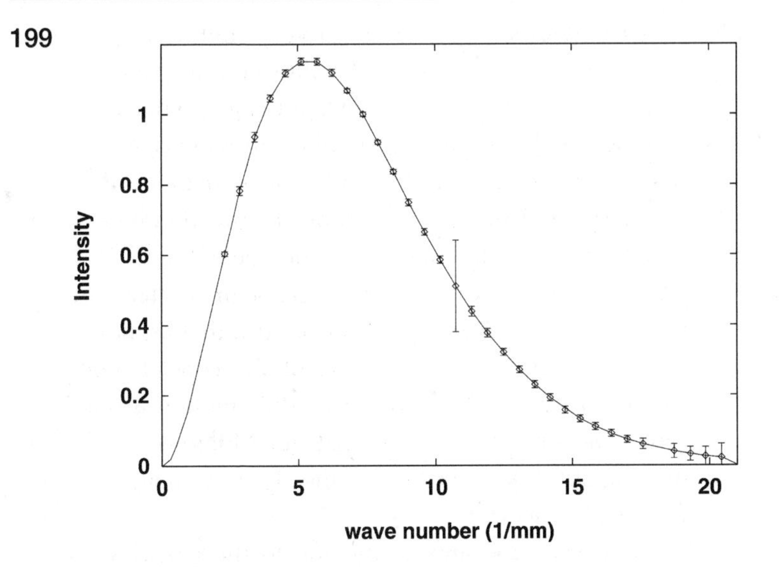

Figure 10.2. Spectrum of the cosmic background radiation as measured by the COBE satellite fitted with a blackbody curve for $T = 2.7277°K$. The "wave number" is defined in terms of the wavelength λ as $k = 2\pi/\lambda$. Note the error bars on the data have been enlarged by a factor of 400. See Fixsen et al. 1996.

to shorter wavelengths. (That's why the coils of your toaster start out "red hot" and then turn "yellow hot" after they heat up some more.) In the case of the cosmic background radiation, the fit to the spectrum yielded $T = 2.7277°K$, a value quite close to what theory had earlier predicted. (A higher temperature would shift the curve in the figure to the right, since the quantity plotted on the x-axis is proportional to the *reciprocal* of the wavelength.)

The background radiation arriving from different directions is almost perfectly identical in amount and temperature, or in other words the radiation is nearly *isotropic*—but not exactly so. Very accurate measurements of the background radiation made in 1989 by the COBE (Cosmic Background Explorer) satellite show that its frequency (and hence its temperature) are both very slightly higher by a mere 0.0033°K in one particular direction in space.

This anisotropy (lack of isotropy) is just what would be expected if the Earth were moving through space in that particular direction, and the radiation were blueshifted as a result. (It's redshifted in the opposite direction in space by exactly the same amount.) From the size of the redshift and blueshift we can calculate that the solar system must be moving through a sea of background radiation at a speed of 370 km/sec in the direction of the constellation Leo. (Coupling this motion with the sun's motion around the center of our galaxy means that our Milky Way galaxy is moving at 600 km/sec toward the Hydra-Centaurus supercluster of galaxies.)

Aside from the anisotropy due to the galaxy's motion through space, it was not until measurements reached the accuracy of one part in a hundred thousand that scientists begin to observe nonuniformities in the background radiation or, in other words, patches of the sky where the radiation is ever so slightly warmer or cooler by a mere 0.00003°K (see fig. 10.3).

Even those tiny nonuniformities represent an important piece of support for the big bang theory, however. In studying the background radiation, we are really looking back in time to when the universe was a mere 300,000 years old (about 0.003 percent of its present age). When we look back in time and see the small nonuniformities in the background radiation temperature, we are also seeing the nonuniformities in the distribution of matter. This is so because concentrations of matter cause the radiation to have a gravitational redshift and appear slightly cooler. If no such tiny nonuniformities in the matter distribution had been present at early times, the big bang theory would have been in big trouble, because it would be difficult to explain how the universe evolved to its present highly clumpy state without some small nonuniformities at the beginning to act as "seeds."

Light Element Abundances

Aside from Hubble's law and the cosmic background radiation, there is still another line of evidence that supports the

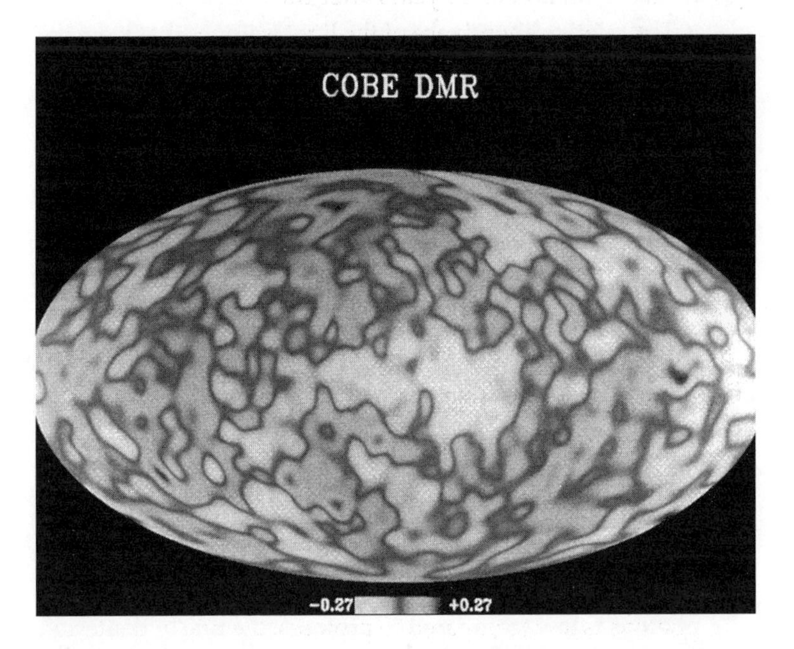

Figure 10.3. A view of the universe as seen in microwave wavelengths, with a great amplification in small temperature nonuniformities. Data are from the COBE satellite (courtesy NASA/Goddard Space Flight Center). The temperature nonuniformities indicated by different gray-level shadings are at the level of one part in 100,000. Positions in the oval correspond to directions in galactic coordinates, with a horizontal line through the middle of the figure corresponding to the plane of the Milky Way galaxy.

 big bang theory, namely, the observed relative abundances of isotopes of the light (low mass) elements. It is believed that the nuclei of these elements (other than ^{1}H) were formed from nuclear reactions inside the primordial fireball, and they were created 0.01 seconds after the big bang. ^{1}H, consisting of a single proton, was created even closer in time to the big bang. In contrast to the light elements, nuclei of heavier elements are believed to have been created or "cooked up" in nuclear reactions inside stars at much later times—a million or so years after the big bang.

The relative abundance of the light isotopes that were created in the primordial fireball depends on the relative density of radiation (photons) and on the matter that was present. To see why this ratio is important, let's consider in particular the nucleus ^{2}H—the isotope of hydrogen known as deuterium, which consists of a bound proton and neutron. The neutron and proton in deuterium are bound together fairly loosely (for a nucleus), and hence their bond can be easily broken in nuclear reactions. Inside the primordial fireball, deuterium nuclei could be created by a collision between two protons: $p + p \rightarrow {}^{2}\mathrm{H} + e^{+}$, but they could also be destroyed by the absorption of a gamma ray photon: $\gamma + {}^{2}\mathrm{H} \rightarrow n + p$.

As a result, the amount of deuterium produced in the fireball depends on an equilibrium between the creation and destruction processes. For example, if the density of photons is low (compared to protons), the rate of deuterium creation exceeds the destruction rate, and, hence, the amount of deuterium would increase. At some abundance of deuterium (the equilibrium abundance), the two creation and destruction rates would become equal. We would expect that the equilibrium abundance of deuterium would decline, the higher the relative density of photons compared to protons, because photons are responsible for the destruction of helium. (It is really no more complex than saying that the amount of housing in the country would have some equilibrium value that occurs when the rate at which new houses are built matches the rate at which old houses are torn down.)

With the other light isotopes, many different nuclear reactions must be considered, but detailed calculations show that their abundances (relative to that of ^{1}H) are all consistent with what is observed in nature, provided that the relative amounts of radiation and matter are also similar to what is observed in nature. (This comparison assumes that we don't count the nonluminous or dark matter. Although unseen, as much as 99 percent of all the matter in the universe is believed to be dark matter.)

Thus, in summary, the big bang theory would seem to be strongly supported by three different types of evidence: the recession of galaxies (Hubble's law), the 2.7°K cosmic background radiation, and the relative cosmic abundances of various light isotopes.

Challenge to the Big Bang Theory

Despite the apparently convincing evidence for the big bang theory, a group of three cosmologists—Geoffrey Burbidge, Fred Hoyle, and Jayant V. Narlikar—have come up with an alternative theory that they believe is at least as consistent with observations and has fewer artificial elements. Their theory of a quasi-steady-state universe, first published in 1993, is an outgrowth of Fred Hoyle's earlier theory of a steady-state universe. In both versions of the steady-state theory, matter is assumed to be continually created over time in minicreation events (little bangs?) rather than all at once as in the big bang theory. (In a pair of back-to-back articles in the April 1999 issue of *Physics Today*, the three proponents of the quasi-steady-state theory debate one of its many critics.)[2]

The difference between the old steady-state theory and the newer quasi-steady-state theory is that the original theory held that the universe was infinitely old and has essentially looked the same at all times and places (hence the name of the theory). The original theory also claimed that newly created matter filled in the empty space created by the universal expansion at a rate just enough to keep the universe always at the same density. The *quasi*-steady-state theory also has no use for a big bang, but it does allow the density of matter in the universe to vary, and in fact it assumes that the universe oscillates in size. Neither version of the steady state theory violates the conservation of energy law, because it is assumed that as new matter and energy are created, it is accompanied by an equal amount of negative energy. The net energy remains constant, therefore, as equal amounts of negative and positive energy are added to the universe at any given point. The ideas of new matter creation and negative energy are not unique to the steady-state theory, nor out of the mainstream. For other ways in which the concept of negative energy is used in physics, see the chapters on time travel and faster-than-light particles.

How would the minicreation process of the steady-state theory actually be observable? According to the theory proponents, one of the sites of minicreation events are black holes. But instead of the conventional view of infalling matter entering the black hole, the steady-staters prefer to describe the process as negative energy emanating from the black hole. Of course, in conventional cosmology, black holes are understood as the remnants of massive stars that have reached the end of their lives. But the proponents of the steady-state theory argue that such an explanation of black holes is implausible, particularly for the supermassive black holes believed to inhabit the centers of galaxies. Steady-staters note that the angular momentum of the swirling matter is so great near the galactic center that matter falling into a black hole would need to have a speed in excess of the speed of light. (However, this argument against a conventional explanation of black hole formation overlooks the removal of excess angular momentum by magnetic fields.)

In the steady-state theory, the rate of creation of new matter is proportional to the density of matter already present at any given point. Therefore, the matter-rich centers of galaxies should be particularly prominent places for new matter creation. According to the theory, this new matter would be expected to include not just ejected subatomic particles, but also large-scale coherent objects, including newly formed galaxies and quasars. Quasars, or quasistellar objects, are compact objects that are observed to have very large redshifts. In the big bang theory, it would be expected (based on Hubble's law) that quasars should all lie at very great distances from us, assuming their redshifts are cosmological in origin—meaning that they are the result of a universal expansion.

But, when quasars were first discovered, it was suggested that at least some of them were actually much closer than their redshift implied. This suggestion was based on a number of cases of quasars having a small angular separation from a nearby galaxy, i.e., being close together on the two-dimensional "dome" of the sky. Although this idea remains highly controversial among mainstream cosmologists, the steady-staters claim that such associations of quasars and nearby galaxies are statistically very significant. In other words, the claim is that some quasars are not just close to nearby galaxies in terms of their two-dimensional position in the sky, but also close to them in three dimensions, having been ejected from the galaxies in minicreation events.

In the conventional big bang theory, the expansion of the universe is described in terms of a time-dependent scale factor $S(t)$, which determines distances such as the separation between galaxies, and the radius of the observable universe. In the big bang theory, the ultimate fate of the universe then depends on the density of matter in the universe. That is because below a certain critical density of about five hydrogen atoms per cubic meter, the universe would keep on expanding forever, since there is just not enough mutual gravitational attraction to halt and reverse the expansion. But above that critical density of matter, the universal expansion will eventually stop and be followed by a collapse, eventually leading to a big crunch and possibly an endless series of expansion and contraction cycles.[2]

In the quasi-steady-state theory, the universe also goes through cycles, but they are not punctuated by big bangs and crunches. Instead, there is a minimum nonzero size of the universe at the start or end of each cycle. Also, superimposed on the cycles there is a slow exponential growth in size of the scale factor at a rate associated with the creation of new matter. During the contraction part of each cycle (when $S(t)$ is decreasing), the negative energy field assumed in the theory causes an outward pressure that halts the contraction at some minimum size, which is taken to be roughly one-sixth of the present size of the universe. In fact, according to the theory, the reversal from a contraction to an expansion is a rather abrupt one, in which the collapsing universe almost seems to rebound from a hard sphere. The

[2] Recent data show that the expansion may actually be accelerating rather than slowing; see endnote [8] of chapter 9.

207 maximum size of the universe at the end of the present cycle is believed to be about twice the present size, and the period of the cycle is about 96 billion years. (The numerical values of all these parameters are chosen in order to make the theory best agree with observations on galactic redshifts and other data.)

You might wonder why three mavericks would bother to come up with such a strange theory when the big bang theory seems to fit the observations so well. One of their justifications is an interesting numerical coincidence involving the cosmic background radiation, considered by many to be the strongest piece of evidence for the big bang. They note that the observed energy density of this radiation, 4.18×10^{-13} erg/cm^3 is quite close to 4.37×10^{-13} erg/cm^3—the value you would expect if this radiation were the result not of a big bang, but rather the product of hydrogen burning in stars.

Thus, according to the three mavericks, a main prop supporting the big bang theory, the background radiation, in fact may have a noncosmic alternative origin. Over time, such stellar burning should produce just the observed density of radiation that is found. You might object that the radiation emitted from various stars has different temperatures characteristic of the particular stars, so it would not wind up having the observed perfect blackbody spectrum that characterizes a body at some fixed temperature. But the quasi-steady-state theory can also explain how the background radiation gets thermalized (i.e., how it winds up with a perfect blackbody spectrum).

During the oscillation cycle, when the size or scale of the universe is near its minimum (roughly one-sixth the present size), the volume would be $1/6^3 = 1/216$ as great, and hence the density of matter would be roughly two hundred times greater than now. In this much denser state, the universe would be much more opaque to light than at present. As a result of the thorough mixing of radiation and matter, both can attain a common uniform temperature or become thermalized. That still leaves the problem of how the radiation goes from a temperature of perhaps 6000 °K, characteristic of visible light, down to a temperature of 3 °K, characteristic of microwave or infrared photons. But the steady-staters have an answer for that problem too: the scattering of light by carbon "whiskers."

Carbon whiskers are threadlike strands that form when carbon vapor condenses. Visible light scattering off these whiskers can get converted to far infrared or microwave wavelengths. According to calculations, there is enough carbon ejected by exploding supernovas to fill space with the needed amount of carbon whiskers, and that could convert scattered visible light to microwaves. Thus, the steady-staters believe that they can account for the uniformity, temperature, and blackbody spectrum shape of the background radiation without requiring a big bang. They even claim to account for the minuscule variations in temperature that have been observed in the background radiation, based on the clustering of matter found in rich clusters of galaxies.

What about the other evidence supporting the big bang theory, such as the relative abundance of the light elements? The steady-staters believe that these can be explained based on nuclear reactions taking place in stars rather than in a primordial fireball. Originally, it was thought that all but eight of the 320 known nuclear isotopes could be accounted for by nuclear processes taking place in stars. Over the years, that list of eight has been narrowed down, and now possible stellar origins have been proposed for all but deuterium. The steady-staters believe that, eventually, it will be shown that this one holdout can also be explained as having a stellar origin.

209
How Good Is the Case against the Big Bang?

Since both the big bang theory and the quasi-steady-state theory claim to explain the observable data, the question is, which does a better job of explaining the data, and which is the theory with fewer ad hoc assumptions? As we shall see, the case against the big bang theory is not terribly strong. Steady-staters claim that they can explain the creation of new matter with their negative energy field, whereas big-bangers merely postulate a big bang in which new matter came from "somewhere else." But that difference seems not quite as stark as the steady-staters claim; it seems more a matter of philosophical preference. In fact, the quasi-steady-state theory, with its exponentially growing scale factor, also assumes that the universe at one time had a very tiny size, just as in the case of the big bang theory—it just didn't start out with tremendously high temperatures, according to the quasi-steady-state theory.

The steady-staters point to the numerical coincidence between the observed density of background radiation and what would be predicted if all that radiation came from hydrogen burning in stars rather than from a big bang. Either this agreement is simply a coincidence, or else the background radiation can be explained as arising from hydrogen burning in stars, and it is not a remnant of the big bang. Obviously, big-bangers choose to believe the agreement is strictly a coincidence. In support of the cosmological origin of the background radiation, they note that the observed features of the background radiation—both its temperature and extraordinarily accurate blackbody shape—are explained quite naturally without free parameters in the big bang theory. In fact, the 3°K background radiation was found *after* the big bang theory had predicted it, while the steady-state theory had to evolve into the quasi-steady-state theory before it could account for some of the properties of this background radiation. Moreover, the quasi-steady-state theory needs to make certain assumptions regarding the nature of the universal scale factor oscillations, as well as the abundance and scattering power and density of hypothetical carbon whiskers, in order to produce thermalization of the background radiation. Even then, the theory is apparently unable to predict accurately the background radiation temperature.

The same could be said about the abundances of the light elements. Nuclear synthesis in the primordial fireball from the big bang is assumed to take place through a series of known reactions at known temperatures. These known reactions allow some light element abundances to be very precisely predicted in the big bang theory with only one free parameter—the ratio of the densities of matter and radiation. In contrast, the steady-staters claim that these light elements arise from hydrogen burning in stars rather than in a cosmic fireball. Given the large range of stellar populations, steady-state theorists apparently have not attempted to explain the actual observed abundances, and they are entirely unable to account for the presence of deuterium, which is consumed and not produced in stellar interiors.

A major problem with the quasi-steady-state theory is that it simultaneously tries to explain too much and too little. Given the idea of minicreation events, the theory could have remained silent as to where in the universe new matter is being created, but, instead, its proponents chose two classes of objects—black holes and quasars—as the manifestations of this process. In rejecting the conventional explanation of black holes as the end state of collapsed massive stars, the theory creates a puzzle as to what *else* could become of massive stars once they have exhausted their fuel

211 and are no longer able to resist the inward pull of gravity. The theory also fails to explain anything else about black holes (such as the expected spectrum of their sizes) that might show that its interpretation of their origin is better than the conventional one.

Similarly, the quasi-steady-state theory postulates that sometimes minicreation events result in the production of large-scale organized structures, without explaining how that could occur. Quasars are identified as these hypothetical structures, because it is claimed that some quasars appear to have been ejected from nearby galaxies. In order to support this claim, the theory needs to invoke the idea that these nearby quasars have intrinsic redshifts that arise not from their motion, but from some unknown mechanism. Yet, simultaneously, the theory also acknowledges that most quasars are at cosmological distances from us. The strongest evidence for the claim that some quasars are relatively close-by is the number of cases where quasars have a small angular separation from some nearby galaxy, but could such associations not arise by chance?

According to several articles written by Burbidge et al., the associations of quasars and nearby galaxies occur far more often than could be explained by chance,[3] yet most studies done by other researchers fail to find any statistically significant result, raising the question of selection biases in the sample used. The selection bias issue is a matter of considerable concern if the observer, who is looking for quasars having a small angular separation from nearby galaxies, is a strong believer that these associations are real. One study by Webster et al. avoided the selection bias issue by using an automated machine to scan astronomical photographs that chose candidate quasars using objective criteria.[4] They found that, of 296 quasars identified, 11 were closely aligned with foreground galaxies. Given the spatial distribution of galaxies, they estimated that such an occurrence would have only a 0.01 percent probability of occurring by chance.

Nevertheless, this finding does not necessarily mean that some quasars actually lie at the same distances as nearby galaxies rather than at the cosmological distances that their large redshifts would seem to imply. The phenomenon of gravitational lensing offers one way to explain the apparent association without a real physical association in three dimensions. According to general relativity, light rays are bent by gravitational fields of massive bodies. Therefore, a massive body acts like a lens that can focus and distort the image of a more distant body seen from Earth. The effect of this focusing action would be to make more distant bodies (the quasars in this case) appear brighter and be more detectable if they happen to lie near the line of sight of the gravitational lens. So, the question comes down to whether there are enough gravitational lenses out there to explain the number of cases in which quasars seem to be associated with nearby galaxies. That number would need to be considerably greater than the number of visible stars.[5] Could there be enough dark matter in the form of dim stars or black holes to provide the needed number of gravitational lenses? Many astronomers are doubtful, which leaves still unsettled the question of whether some quasars are, in fact, closer than cosmological distances.

Even in the event that it could be definitely shown that some quasars with large redshifts are physically associated with nearby galaxies, that finding would not, of itself, establish that the quasars have been ejected by the galaxies. In fact, if the ejected quasars have their large redshifts as a result of a motion-induced Doppler shift, we would expect to see some quasars with large blueshifts (if they happen to be approaching us), and that has never been seen. This absence of blueshifts is why the steady-state theorists need to invent the concept of an intrinsic redshift of unknown origin. This ad hoc assumption, made in order to account for the idea of ballistically ejected quasars—an idea that is not central to the theory—seems questionable, and it does not inspire great confidence in the quasi-steady-state theory. In summary, it would appear that the alternative to the big bang theory does a considerably poorer job in explaining the observations and requires more ad hoc assumptions.

There are, nevertheless, some mysterious observations that are unexplained in the conventional big bang cosmology. Perhaps the most important one is the near absence of antimatter in the universe. At the extremely high temperatures present shortly after the big bang, matter and antimatter should have been present in equal amounts, so it is unclear what happened to the antimatter.[3] One variation of the big bang theory proposed by J. Richard Gott, however, is able to account for the missing antimatter.[6]

According to Gott's time-symmetric theory, three universes were created in the big bang: universe I, dominated by matter and evolving into the future; universe II, a time-reversed universe dominated by antimatter; and, finally, universe III, a faster-than-light tachyon universe. The three universes correspond to the threefold division of spacetime based on the past and future light cones (see fig. 8.1). Imagine that the big bang occurs at the origin of spacetime—the point where the past and future light cones meet—and that matter travels outward from that point in *all* possible directions in spacetime. The resulting worldlines include ordi-

[3] It is possible that initially there was slightly more matter than antimatter present, and our universe is the remnant left over after most of the matter was annihilated by antimatter. But in that case, much more radiation should be present than is observed.

 nary matter moving forward in time (inside the future light cone, antimatter moving backward in time (inside the past light cone), and faster-than-light tachyons (outside both light cones). Gott's time-symmetric theory can account for the near absence of antimatter, in our universe, since the antimatter expanding backward in time from the big bang could be interpreted as a matter universe collapsing to the big bang. These three universes fill the three regions of the light cone centered on the big bang, and they represent the most general solution to Einstein's general relativity field equations. Gott's version of the big bang theory treats matter, antimatter and tachyons on an equal footing, and it is consistent with observations. It would receive additional support if tachyons are confirmed as being real particles (see previous chapter).

My rating for the idea that the big bang never happened is 3 cuckoos.

11 Epilogue

I HOPE you have enjoyed reading this book as much as I've enjoyed writing it. If you did, you might want to send me money or, alternatively, to check out some of my other books listed in the bibliography. There is a reasonable likelihood I may be writing a sequel—"More Crazy Ideas in Science That Might Be True." Accordingly, I'd like to hear from you if you have suggestions for topics. I had thought of asking that you refrain from sending me pet theories or "nutty" ideas (as distinct from crazy ones), but if you got this far in the book you are likely to be in possession of most of your marbles. And if I'm mistaken on that score, any request on my part that you avoid sending me nutty ideas would be likely to be ignored. Please send your non-nutty suggestions to me at rehrlich@gmu.edu.

At the end of this epilogue you can find a summary of my subjective ratings for the likelihood that each of the nine ideas in this book is incorrect. As I've noted, these judgments are entirely subjective, and someone with more intimate knowledge than I of any one of these subjects might well disagree with my ratings. Of course, having a detailed knowledge of a given subject can sometimes be a handicap instead of a benefit, since experts may be more used to thinking in terms of the conventional paradigm for their field.

In the introduction to this book, I suggested a number of ways to tell if a crazy idea might be true. Some of these criteria relate to the proposer of the idea, including his or her qualifications, agenda, and degree of objectivity. (One can imagine any number of reasons someone might propose a crazy idea that is nonsense, including, ignorance, self-delusion, outright blunders, a desire for fame and fortune, or an earnest wish to save the planet.) Other criteria for spotting crazy ideas that are likely to be false relate to the manner in which the idea is presented: the care in the use of statistics, the number of references to the work of others, and the openness in presenting data and methods of analysis.

Still other criteria relate to the idea itself—most importantly, the question of its "falsifiability." Any theory that deserves the label "scientific" must be capable of being tested and proven wrong, and the idea's proposer must tell us how this could be done. A certain amount of refinement of the theory (fudging?) in response to new and changing evidence may be tolerated, but any theory that can be revised in order to make it fit new evidence (and can therefore never be proven wrong), is outside the realm of science—even if it is dressed up with lots of impressive scientific terminology. At the other extreme, when a theory is very specific in its predictions, we can be that much more confident in it if those predictions should come true.

Other tests of a theory would include the range in its explanatory power—though we should be suspicious of theories that claim to explain virtually everything—and the extent to which the theory agrees with our common sense. The latter criterion can be especially tricky, however, since some crazy ideas that turned out to be great revolutionary breakthroughs, such as quantum theory, seem to be at odds with our common sense, even now, roughly seventy years after its creation. The criterion of simplicity ("Occam's razor") can be considered one aspect of common sense. For example, other things being equal, it would seem simpler to diagnose a bump on your arm as a pimple, rather than as an implanted device put there by aliens who abducted you during the night. On the other hand, the essence of simplicity does depend on your belief system. If you happen to be among millions of Americans who suffer from the delusion that alien abductions are commonplace events, the idea of an implanted device may seem just as "simple" an explanation for the bump as a pimple.

Throughout the history of science, there have been any number of cases of ideas once thought to be crazy that later turned out to be true, and many more cases where they turned out to be false. Science is always a work in progress, and today's truths always have a provisional nature. In order to make cumulative progress, science must demand that new crazy ideas pass a very high threshold before they can displace existing ideas that have proven their worth. The acid test of the worth of any crazy idea must be the empirical evidence for and against it, rather than our beliefs. Moreover, crazy ideas are not shown to be true simply by finding the evidence which supports them. The best way of proving the correctness of a crazy idea is to do everything possible to show it to be false—and fail.

Some proponents of crazy ideas claim to be skeptics who were forced to accept the idea because of the evidence, but their self-described skepticism sometimes doesn't ring true because of the way they interpret the evidence. In weighing evidence for and against an idea, it is most important to use the same standards on each side of the argument. If you believe, for example, that violent video games partly explain the rise in youth violence, you should be prepared to rethink your view when faced with the statistic that teen violence has actually dropped significantly during the time when violent video games became prevalent. Actually, such parallel time trends offer little support on one side or the other, since a correlation between two trends, by itself, does not imply that one trend caused the other ("correlation does not imply causation"). Another useful saying to bear in mind when weighing controversial claims is that "the absence of evidence is not evidence of absence." To take one familiar example, the failure to find a specific natural (or man-made) explanation for a particular unidentified flying object (UFO) does not imply that no such explanation exists.

As indicated by the preceding examples, the criteria applied here for testing the truth or falsity of controversial scientific claims should also be useful in everyday life. There are any number of important issues relating to human health, technology, and the environment for which conflicting evidence can be found: Are cell phones dangerous? Can wearing magnets promote health? Is global warming likely to occur, and how harmful will it be? I hope that after reading this book you may be in a better position to evaluate the claims made by experts on each side, and resolve such issues for yourself.

You will also need to decide if you want to be a supporter of a cause or a truth seeker. (If you have already decided that global warming represents an impending catastrophe for humanity, you are unlikely to wish to weigh dispassionately arguments on each side of the debate.) Unlike supporters of causes, truth seekers must, above all, always be modest in what they can know with certainty, and remain open to new evidence—both in support of their view and opposed to it. Despite what you think you know, the next crazy idea you come across could prove to be true. But it would also be useful to remember Carl Sagan's aphorism that extraordinary claims require extraordinary evidence to support them.

Finally, in assessing crazy ideas, don't forget that seeing should not always be believing: while nature doesn't lie or pull tricks on us, our fellow humans sometimes do. It is interesting, for example, that virtually all claims of paranor-

219 mal abilities involve feats, such as spoon bending, which can be (and have been) easily reproduced by professional magicians. Of course, that doesn't necessarily imply that trickery (rather than paranormal ability) was the explanation in any given case, but which explanation of the two do you think is simpler?

The table below shows my ratings for the nine ideas discussed in this book. I've used a subjective rating scheme that goes like this: Zero cuckoos means "why not?" One cuckoo means "probably not true, but who knows?" Two cuckoos means "very likely not true." Three cuckoos means "almost certainly not true." And four cuckoos means "certainly false."

Crazy Idea	*Rating*
More guns means less crime	3 cuckoos
AIDS is not caused by HIV	3 cuckoos
Sun exposure is beneficial	0 cuckoos
Low doses of nuclear radiation are beneficial	1 cuckoo
The solar system has two suns	2 cuckoos
Oil, coal, and gas have abiogenic origins	0 cuckoos
Time travel is possible	2 cuckoos
Faster-than-light particles exist	0 cuckoos
There was no big bang	3 cuckoos

Notes to the Chapters

Chapter 2. More Guns Means Less Crime

[1] The low estimate is cited by Philip Cook based on the National Crime Victimization Survey Reports; the high estimate is based on fifteen national polls. Both are cited in John Lott Jr., *More Guns, Less Crime* (Chicago: University of Chicago Press, 1998), 11.

[2] Based on national surveys, cited in ibid., 3.

[3] James D. Wright and Peter Rossi, *Armed and Considered Dangerous: A Survey of Felons and Their Firearms* (Hawthorn, NY: Aldine de Gruyter, 1986).

[4] See Lott, *More Guns*, 5, for references.

[5] U.S. Department of Justice, FBI Staff, Uniform Crime Reports (Washington, DC: U.S. Government Printing Office, 1992).

[6] Editorial, *Cincinnati Enquirer*, Jan. 23, 1996, A8.

[7] David McDowall, Colin Loftin, and Brian Wiersema, Easing concealed firearm laws: Effects on homicide in three states, *J. Crim. Law and Criminology* 86 (Fall 1995): 193–206.

[8] Arthur L. Kellerman et al., Gun ownership as a risk factor for homicide in the home, N. Engl. Med. Oct. 7, 1993, 1084–91.

[9] Cited in Lott, *More Guns*, 41.

[10] Cited in *The Economist*, Arms and the man, July 3, 1999. Bellesiles's research will be published in a forthcoming book.

[14] J. R. Lott Jr. and M. Landes, Multiple victim public shootings, bombings and right-to-carry concealed handgun laws: Contrasting private and public law enforcement, John M. Olin Law and Economics Working Paper no. 73 (2D series).

Chapter 3. AIDS Is Not Caused by HIV

[1] P. H. Duesberg, *Inventing the AIDS Virus* (Washington, DC: Regnery Publishing, 1996).

[2] P. H. Duesberg, *Stretching the Germ Theory beyond Its Limits* (Berkeley, CA: North Atlantic Books, 1996).

[3] P. H. Duesberg and J. Yiamouyiannis, *AIDS: The Good News Is HIV Doesn't Cause It* (Dordrecht, The Netherlands: Health Action Press, 1995).

[4] P. H. Duesberg, ed., *AIDS: Virus or Drug Induced?* (Dordrecht, The Netherlands: Kluwer Academic Publishers, 1996).

[5] Centers for Disease Control, *Morbidity and Mortality Weekly Report* (*MMWR*), 31, 507–14 (1982); *MMWR*, 34, 373–75 (1985); *MMWR*, 36, 3S–15S (1987); *MMWR*, 41, 1–19 (1992).

[6] A list of Peter Duesberg's publications can be found on the Web site: www.virusmyth.com/aids/index/pduesberg.htm.

[7] Duesberg, *Inventing the AIDS Virus*, 402.

[8] Centers for Disease Control, *MMWR*, 30, 250–52 (1981); *MMWR*, 30, 305–8 (1981).

[9] C.K.O. Williams et al., AIDS-associated cancers, in M. Essex et al., eds., *AIDS in Africa* (New York: Raven Press, 1994), 325–71.

[10] J. J. Goedert et al., Amyl nitrite may alter T lymphocytes in homosexual men, *Lancet*, 1(8269), 412–16 (1982).

[11] R. D. deShazo et al., An immunologic evaluation of hemophiliac patients and their wives: Relationships to the acquired immunodeficiency syndrome, *Ann. Intern. Med.*, 99(2), 159–64 (1983).

[12] R. C. Gallo and L. Montagnier, The chronology of AIDS research, *Nature*, 326(6112), 435–36 (1987).

[13] M. G. Sarngadharan, M. Popovic, L. Bruch, J. Schupbach, and R. C. Gallo, Antibodies reactive with human T-lymphotropic retroviruses (HTLV-III) in the serum of patients with AIDS, *Science*, 224(4648), 506–8 (1984).

[14] R. Cheingsong-Popov et al., Prevalence of antibody to human Y-lymphotropic virus type III in AIDS and AIDS-risk patients in Britain, *Lancet*, 2(8401), 477–80 (1984).

[15] R. C. Gallo and M. S. Reitz Jr., Human retroviruses and adult T-cell leukemia-lymphoma, *J. Natl. Cancer Inst.*, 69(6), 1209–14 (1982).

[16] G. Pantaleo, C. Graziosi, and A. S. Fauci, The immunopathogenesis of human immunodeficiency virus infection, *N. Engl. J. Med.*, 328(5), 327–35 (1993).

223 [17] A. S. Fauci, Multifactorial nature of human immunodeficiency virus disease: Implications for therapy, *Science*, 262(3136), 1011–18 (1993).

[18] G. F. Lemp et al., Projections of AIDS morbidity and mortality in San Francisco, *JAMA*, 263(11), 1497–1501 (1990).

[19] Centers for Disease Control, Revised classification system for HIV infection and expanded surveillance case definition for AIDS among adolescents and adults, *MMWR*, 41, 1–19 (1992).

[20] K. Saksela, C. Stevens, P. Rubinstein, and D. Baltimore, Human immunodeficiency virus type 1 mRNA expression in peripheral blood cells predicts disease progression independent of the number of CD4+ lymphocytes, *Proc. Natl. Acad. Sci. USA*, 91, (3), 1104–8 (1994).

[21] Duesberg, *Inventing the AIDS Virus*, 175.

[22] Ibid., 187.

[23] P. Duesberg and D. Rasnick, The AIDS dilemma: Drug diseases blamed on a passenger virus, *Genetica*, 104, 85–132 (1998).

[24] *Washington Post*, April 30, 2000.

[25] CDC data presented at the 1999 National Prevention Conference, Atlanta, August 29–September 1, 1999.

[26] UNAIDS, AIDS epidemic update, 1999.

[27] Duesberg, *Inventing the AIDS Virus*, 290.

[28] World Health Organization, *The Current Global Situation of the HIV/AIDS Pandemic*, January 3, 1995.

[29] *Washington Post*, May 4, 2000.

[30] Duesberg, *Inventing the AIDS Virus*, 189.

[31] As one of his definitions of AIDS, Duesberg uses, for example, the ratio of T4 to T8 cells being less than one. See appendix C of *Inventing the AIDS Virus*.

[32] D. K. Smith, J. J. Neal, and S. D. Holmberg, Unexplained opportunistic infections and CD4+ T-lymphocytopenia without HIV infection: An investigation of cases in the United States, *N. Engl. J. Med.*, 328(6), 373–79 (1993).

[33] S. Kwock et al., Identification of human immunodeficiency virus sequences by using in vitro enzymatic amplification and oligomer cleavage detection, *J. Virol.*, 61(5), 1690–94 (1987).

[34] P. H. Duesberg, Results fall short for AIDS theory, *Insight*, Feb. 14, 1994, 27–29.

[35] Duesberg, *Inventing the AIDS Virus*, 274.

[36] P. H. Duesberg, Retroviruses as carcinogens and pathogens: Expectations and reality, *Cancer Res.*, 47(5), 1199–220 (1987).

[37] W. Blattner, R. C. Gallo, and H. M. Temin, HIV causes AIDS, *Science*, 241(4865), 515–16 (1988).

[38] Centers for Disease Control, HIV/AIDS Surveillance Report, 1994 year-end edition, 6(2) (1995).

[39] Duesberg, *Inventing the AIDS Virus*, 182.

[40] Ibid., 183.

[41] S. W. Barnett, K. K. Murthy, B. G. Herndier, and J. A. Levy, An AIDS-like condition induced in baboons by HIV-2, *Science*, 266, 642–46 (1994).

[42] B. L. Evatt, E. D. Gomperts, J. S. McDougal, and R. B. Ramsey, Coincidental appearance of LAV/HTLV-III antibodies in hemophiliacs and the onset of the AIDS epidemic, *N. Engl. J. Med.*, 312(8), 483–86 (1985); A. S. Fauci, The human immunodeficiency virus: Infectivity and mechanisms of pathogenesis, *Science*, 239(4840), 617–22 (1988).

[43] Centers for Disease Control, Persistent, generalized lymphadenopathy among homosexual males, *MMWR*, 31, 249–52 (1982).

[44] S. C. Darby et al., Mortality before and after HIV infection in the complete UK population of haemophiliacs, *Nature*, 377(7), 79–82 (1995).

[45] J. K. Kreiss et al., Nontransmission on T-cell subset abnormalities from hemophiliacs to their spouses, *JAMA*, 251, 1450–54 (1984), cited in Duesberg, *Inventing the AIDS Virus*.

[46] Centers for Disease Control, HIV/AIDS Surveillance Report through December 1998, 10(2) (1998).

[47] B. Z. Katz, Natural history and clinical management of the infant newborn to a mother infected with human immunodeficiency virus, *Semin. Perinatal.*, 13(1), 27–34 (1989).

[48] C. L. Park, H. Streicher, and R. Rothberg, Transmission of human immunodeficiency virus to only one dizygotic twin, *J. Clin. Microbiol.*, 25(6), 1119–21 (1987).

[49] H. W. Jaffe et al., The acquired immunodeficiency syndrome in a cohort of homosexual men: A six-year follow-up study, *Ann. Intern. Med.*, 103(2), 210–14 (1985).

225 [50] Institute of Medicine, National Academy of Sciences, *Confronting AIDS: Directions for Public Health, Health Care and Research* (Washington, DC: National Academy Press, 1986).

[51] M. S. Ascher et al., Does drug use cause AIDS? *Nature,* 362(6416), 103–4 (1993).

[52] M. T. Schechter et al., HIV-1 and the aetiology of AIDS, *Lancet,* 341, 658–59 (1993); P. J. Veugelers et al., Determinants of HIV disease progression among homosexual men registered in the Tricontinental Seroconverter Study, *Am. J. Epidemiol.,* 15, 140(8), 747–58 (1994).

[53] J. J. Goedert et al., Risks of immunodeficiency, AIDS, and death related to purity of factor VIII concentrate, *Lancet,* 344, 791–92 (1994), cited in Duesberg, *Inventing the AIDS Virus.*

[54] M. A. Sande et al., Antiretroviral therapy for adult HIV-infected patients: Recommendations for a state-of-the-art conference, *JAMA,* 270(21), 2583–89 (1993).

[55] M. A. Fischl et al., The safety and efficacy of zidovudine (AZT) in the treatment of patients with AIDS and AIDS-related complex: A double-blind, placebo-controlled trial, *N. Engl. J. Med.,* 317(4), 185–91 (1987).

[56] G. X. McLeod and S. M. Hammer, Zidovudine: Five years later, *Ann. Int. Med.,* 117(6), 487–501 (1992).

[57] See fig. 2(A) in Duesberg and Rasnick, The AIDS dilemma; data from Centers for Disease Control (1995).

Chapter 4. Sun Exposure Is Beneficial

[1] J. Adami, M. Frisch, J. Yuen, B. Glimelius, and M. Melbye, Evidence of an association between non-Hodgkin's lymphoma and skin cancer, *Br. Med. J.,* 310, 1491–95 (1995).

[2] International Agency for Research on Cancer, *IARC Monographs on the Evaluation of Carcinogenic Risks to Humans: Ultraviolet Radiation,* vol. 55 (Lyon: IARC, 1992).

[3] J. R. M. Kunz and A. J. Finkel, eds., *The Family Medical Guide of the American Medical Association* (New York: Random House, 1987.

226 [4] J. M. Elwood and J. Jopson, Melanoma and sun exposure: An overview of published studies, *Int. J. Cancer*, 73, 198–203 (1997).

[5] The two-standard-deviation range quoted in ibid. is 1.54–1.90.

[6] The two-standard-deviation range quoted in ibid. is 0.77–0.96.

[7] J. M. Elwood, G., B. Gallagher, G. B. Hill, and J.C.G. Pearson, Cutaneous melanoma in relation to intermittent and constant sun exposure—the western Canada melanoma study, *Int. J. Cancer*, 35, 427–33 (1985).

[8] For a justification, see, for example, A. H. Rosenfeld, *Ann. Rev. Nucl. Sci.*, 25, 555 (1975).

[9] A. J. McMichael and A. J. Hall, Does immunosuppressive ultraviolet radiation explain the latitude gradient for multiple sclerosis? *Epidemiology*, 8, 642–45 (1997).

[10] A. E. Evans et al., Autres pays, autres coeurs? Dietary patterns, risk factors and ischaemic heart disease in Belfast and Toulouse (WHO MONICA project), *Q. J. Med.*, 88, 469–77 (1995).

[11] Toulouse is 12 degrees closer to the equator in latitude than Belfast, which would result in twice the annual UV exposure, if the extent of cloud cover were the same. In fact, because Toulouse is much sunnier than Belfast, the difference is much larger than that.

[12] D. S. Grimes, E. Hindle, and T. Dyer, Sunlight, cholesterol and coronary heart disease, *Q. J. Med.*, 89, 579–89 (1996).

[13] R. Scragg et al., Myocardial infarction is inversely associated with plasma 25-hydroxyvitamin D3 levels: A community-based study, *Int. J. Epidemiol.*, 19, 559–63 (1990).

[14] R. Fabsitz and M. Feinleib, Geographic patterns in country mortality rates from cardiovascular diseases, *Am. J. Epidemiol.*, 111, 315–28 (1980).

[15] E. A. Mortimer, M. D. Monson, and B. MacMahon, Reduction in mortality from coronary heart disease in men residing at high altitude, *N. Engl. J. Med.*, 296, 581–85 (1977).

[16] A. W. Voors, and W. D. Johnson, Altitude and arteriosclerotic heart disease mortality in white residents of 99 of the 100 largest cities in the United States, *J. Chronic Dis.*, 32, 157–62 (1979).

[17] P. H. Kutchenreuter, *N.Y. Acad. Sci. Trans.*, 22, 126 (1959).

227 [18] M. F. Muldoon, S. B. Manuck, and K. A. Matthews, Lowering cholesterol concentrations and mortality: A quantitative review of primary prevention trials, *Br. Med. J.*, 301, 309–14 (1990).

[19] U. Ravnskov, Cholesterol lowering trials in coronary heart disease: Frequency of citation and outcome, *Br. Med. J.*, 305, 15–19 (1992).

[20] J. Elford, A. N. Phillips, A. G. Thomson, and A. G. Shaper, Migration and geographic variations in ischaemic heart disease in Great Britain, *Lancet*, 1, 343–46 (1989).

[21] P. M. McKeigue and M. G. Marmot, Mortality from coronary heart disease in Asian communities in London, *Br. Med. J.* (1988).

[22] L. O. Hughes, U. .Raval, and E. B. Raftery, First myocardial infarctions in Asian and white men, *Br. Med. J.*, 298, 1345–50 (1989).

[23] R. Smith, Asian rickets and osteomalacia, *Q. J. Med.*, 76, 899–901 (1990).

[24] P. J. Finch, F.J.C. Millard, J. D. Maxwell, Risk of tuberculosis in immigrant Asians: Culturally acquired immunodeficiency? *Thorax*, 46, 1–5 (1991).

[25] P. J. Finch, L. Ang, J. B. Eastwood, and J. D. Maxwell, Clinical and histological spectrum of osteomalacia among Asians in South London, *Q. J. Med.*, 83, 439–48 (1992).

[26] Although we have mainly focused on the effects on humans, a growing number of studies indicate UV radiation is harmful to acquatic organisms. See, for example, P. Kuhn and H. Browman, Penetration of ultraviolet radiation in the waters of the estuary and gulf of the Saint Lawrence, *Limnol. Oceanogr.*, 44, 710–16 (1999).

Chapter 5. Low Doses of Nuclear Radiation Are Beneficial

[1] Jeffrey R. M. Kunz and Asher J. Finkel, eds., *The American Medical Association Family Medical Guide* (New York: Random House, 1987), 507.

[2] Sohei Kondo, *Health Effects of Low-Level Radiation* (Osaka: Kinki University Press, and Madison, WI: Medical Physics Publishing, 1993).

[3] M. Mine, T. Nakamura, H. Mori, H. Kondo, and S. Okajima, The Current mortality rates of A-bomb survivors in Nagasaki City, *Jpn. J. Public Health*, 28, 337–42 (1981), in Japanese with English abstract, cited in Kunz and Finkel, *Family Medical Guide*, 28.

[4] M. Mine, Y. Okumura, M. Ichimaru, T. Nakamura, and S. Kondo, Apparently beneficial effect of low to intermediate doses of A-bomb radiation on human lifespan, *Int. J. Radiat. Biol.*, 58, 1035–43 (1990), cited in Kunz and Finkel, *Family Medical Guide*, 28.

[5] D. A. Pierce, Y. Shimizu, D. L., Preston, M. Vaeth, and K. Mabuchi, Studies of the mortality of atomic bomb survivors, Report 12, part 1, Cancer: 1950–1990, *Radiat. Res.*, 146, 1–27 (1996), cited in B. L. Cohen, Validity of the linear no-threshold theory of radiation carcinogenesis in the low dose region, *Health Phys.*, 6, 43–61 (1999). This article references five earlier papers on Cohen's radon study.

[6] R. D. Evans, Radium in man, *Health Phys.*, 27, 497–510 (1974), cited in Cohen, see ref. 5.

[7] Cohen, Validity of the theory, 43–61.

[8] N. A. Frigerio, K. F. Eckerman, and R. S. Stower, The Argonne radiological impact program, in *Environmental and Earth Sciences Report ANI/ES-26*, part 1 (Argonne, IL: Argonne National Laboratory), cited in J. H. Fremlin, *Power Production: What Are the Risks*? (Bristol and Boston: Adam Hilger, 1985), 56.

[9] J. H. Lubin and J. D. Boice, Lung cancer risk from residential radon: Meta-analysis of eight epidemiological studies, *J. Nat. Cancer Inst.*, 89, 49–57 (1997).

[10] J. H. Lubin et al., Radon and lung cancer risk: A joint analysis of 11 underground miners studies, National Institutes of Health, NIH pub. 94–3644 (1994).

[11] J. H. Lubin, On the discrepancy between epidemiologic studies in individuals of lung cancer and residential radon and Cohen's ecologic regression, *Health Phys.*, 75, 4–10 and 29–30 (1998).

[12] See B. L. Cohen, Response to Lubin's proposed explanation of our discrepancy, *Health Phys.*, 75, 18–22, (1998), for a derivation of the relationship between r and r'.

229 [13] See, for example, S. Greenland and J. Robbins, Invited commentary: Ecologic studies—biases, misconceptions and counterexamples, *Am. J. Epidemiol.*, 139, 747–60 (1994).

[14] Various arguments are discussed in Cohen, Validity of the theory, which also lists a number of references on the protective effect of a low radiation dose before a high dose is given.

[15] For a discussion of EPA policies and scare tactics, see L. A. Cole, *Element of Risk: The Politics of Radon* (New York: Oxford University Press, 1993), 154, appendix C.

Chapter 6. The Solar System Has Two Suns

[1] L. W. Alvarez, W. Alvarez, F. Asaro, and H. V. Michel, Extraterrestrial causes for the Cretaceous-Tertiary extinction, *Science*, 208, 1095–1108 (1980).

[2] R. P. Turco et al., Nuclear winter: Global consequences of multiple nuclear explosions, *Science*, 222, 1283–92 (1983). For some critical analyses, see R. Ehrlich, *Waging Nuclear Peace: The Technology and Politics of Nuclear Weapons* (Albany: State University of New York Press, 1985).

[3] For a discussion of some of the criticisms of the Alvarez discovery, see W. Alvarez, *T. rex and the Crater of Doom* (Princeton, NJ: Princeton University Press, 1997).

[4] W. Alvarez and R. A. Muller, Evidence from crater ages for periodic impacts on the Earth, *Nature*, 308, 718–20 (1984).

[5] D. M. Raup and J. J. Sepkoski Jr., Periodicity of extinctions in the geologic past, *Proc. Natl. Acad. Sci.*, 81, 801–5 (1984).

[6] J. J. Sepkoski Jr., Periodicity in extinction and the problem of catastrophism in the history of life, *J. Geol. Soc., London*, 146, 7–19 (1989).

[7] M. Davis, P. Hut, and R. A. Muller, Extinction of species by periodic comet showers, *Nature*, 308, 715–17 (1984).

[8] M. R. Rampino and R. B. Stothers, Terrestrial mass extinctions, cometary impacts and the sun's motion perpendicular to the galactic plane, *Nature*, 308, 709–12 (1984); R. D. Schwartz and P. B. James, Periodic mass extinctions and the sun's oscillation about the galactic plane, *Nature*, 308, 712–13 (1984).

230 [9] J. G. Hills, The passage of a "Nemesis"-like object through the planetary system, *Ap. J.*, 90, 1876–82 (1985).

[10] M. R. Rampino, and B. M. Haggerty, The "Shiva Hypothesis": Impacts, mass extinctions, and the galaxy, *Earth, Moon, and Planets: An International Journal of Comparative Planetology*, 72, 441–60 (1996).

[11] See, for example, E. Noma and A. L. Glass, Mass extinction pattern: Result of chance, *Geol. Mag.*, 124 (4), 319–22 (1987), and S. M. Ross, Are mass extinctions really periodic? *Probability in the Engineering and Informational Sciences*, 1, 61–64 (1987).

[12] S. Perlmutter et al., The Berkeley search for a faint stellar companion to the sun, in *Astrophysics of Brown Dwarfs*, ed. M. Kafatos, R. Harrington, and S. Maran (New York: Cambridge University Press, 1986).

Chapter 7. Oil, Coal, and Gas Have Abiogenic Origins

[1] T. Gold, *The Deep Hot Biosphere* (New York: Springer-Verlag, 1999).

[2] T. Gold, *Power from the Earth* (London: J. M. Dent & Sons, 1987).

[3] T. Gold, The deep, hot biosphere, *Proc. Natl. Acad. Sci. USA*, 89, 6045–49 (1992).

[4] T. Gold, The origin of methane in the crust of the Earth, *U.S. Geological Survey Prof. Paper 1570* (1993).

[5] E. B. Chekaliuk, and J. F. Kenney, The Stability of hydrocarbons in the thermodynamic conditions of the Earth, *Proc. Am. Phys. Soc.*, 36(3), 347 (1991).

[6] N. S. Beskrovny and N. I. Tikhomirov, Bitumens in the hydrothermal deposits of Transbaukal, in The genesis of oil and gas, *Izdvo Nedra* (1968), cited in Gold, Origin of methane.

[7] R. Robinson, Duplex origin of petroleum, *Nature*, 199, 113–14 (1963); idem, The origins of petroleum, *Nature*, 212, 1291–95 (1966), cited in Gold, Origin of methane.

[8] N. A. Kudryavtsev, Geological proof of the deep origin of petroleum, *Trudy Vsesoyuz. Neftyan. Nauch.-Issledovatel.*

231 *Geologoraz Vedoch. Inst.*, no. 132, 242–62 (1959), cited in Gold, Origin of methane.

[9] T. Gold, Sweden's Siljan Ring well evaluated, *Oil and Gas J.*, Jan. 14, 1991, 76–78.

[10] See Gold, *The Deep Hot Biosphere*, 26.

[11] F. K. North, Review of Thomas Gold's deep-earth-gas hypothesis, *Energy Explor. Exploit*, 1, 105–10 (1982).

[12] See Gold, *The Deep Hot Biosphere*, 128.

[13] Ibid., 214n.6.

[14] E. M. Galimov, Isotopic composition of carbon in gases of the crust, *Intern. Geol. Rev.*, 11, 10, 1092–1104 (1969), cited in Gold, Origin of methane.

[15] E. M. Galimov and K. A. Kvenvolden, Concentrations and carbon isotopic compositions of CH_4 and CO_2 in gas from sediments of the Blake Outer Ridge, Deep Sea Drilling Project Leg 76, *Initial Reports of the Deep Sea Drilling Project*, 76, 403–7 (1983), cited in Gold, Origin of methane.

[16] M. Schidlowski, R. Eichmann, and C. E. Junge, Precambrian sedimentary carbonates—Carbon and oxygen isotope geochemistry and implications for the terrestrial oxygen budget, *Precambrian Research*, 2, 1–69 (1975), cited in Gold, Origin of methane.

Chapter 8. Time Travel Is Possible

[1] J. Mallinckrodt, "What happens when *at* is greater than *c*? Talk given at the 1999 Summer AAPT meeting. Also see http://www.csupomona.edu/~ajm/myweb/talks/atgtc.pdf.

[2] F. J. Tipler, Rotating cylinders and the possibility of global causality violation, *Phys. Rev.*, D9, 2203–6 (1974).

[3] K. Gödel, An example of a new type of cosmological solution of Einstein's field equations of gravitation, *Rev. Mod. Phys.*, 21, 447–50 (1949).

[4] J. R. Gott, Closed timelike curves produced by pairs of moving cosmic strings: Exact solutions, *Phys. Rev. Lett.*, 66, 1126–29 (1991).

232 [5] M. S. Morris, and K. S. Thorne, Wormholes, time machines, and the weak energy condition, *Phys. Rev. Lett.*, 61, 1446–49 (1988).

[6] A complete discussion and review of the literature of the Casimir effect may be found in Resource Letter CF-1: S. K. Lamoreaux, Casimir force, *Am. J. Phys.*, 67, 850–61 (1999).

[7] S. W. Hawking, Particle creation by black holes, *Comm. in Math. Phys.*, 43, 199–220 (1975).

[8] A. G. Riess, et al., Observational evidence from supernovae for an accelerating universe and a cosmological constant, *Astr. J.*, 116, 1008–38 (1998).

[9] S. Hawking, *Phys. Rev.*, D47, 554–65 (1993).

[10] H. Everett III, Relative state formulation of quantum mechanics, *Rev. Mod. Phys.*, 29, 454–62 (1957).

Chapter 9. Faster-than-Light Particles Exist

[1] A. Einstein, On the electrodynamics of moving bodies, *Ann. Physik*, 17, 891 (1905).

[2] O. M. P. Bilaniuk, V. K. Deshpande, and E.C.G. Sudarshan, "Meta" Relativity, *Am. J. Phys.* 30, 718 (1962).

[3] For a discussion and list of references, see P. J. Nahin, *Time Machines: Time Travel in Physics, Metaphysics, and Science Fiction* (New York: American Institute of Physics Press, 1993).

[4] See, for example, for several negative searches: T. Alvager and M. N. Kreisler, Quest for faster-than-light particles, *Phys. Rev*, 171, 1357 (1968), and C. Baltay, G. Feinberg, N. Yeh, and R. Linsker, Search for uncharged faster-than-light particles, *Phys. Rev.*, D1, 759 (1970). For the one positive report, see R. W. Clay and P. C. Crouch, Possible observations of tachyons associated with extensive air showers. *Nature*, 248, 29 (1974).

[5] For current best estimates on the neutrino masses, see C. Caso et al., Null experiments for neutrino masses, *European Phys. J.*, C3, 1 (1998). Also see http://pdg.lbl.gov.

[6] A. Chodos, V. A. Kostelecky, R. Potting, and E. Gates, Mass bounds for spacelike neutrinos, *Phys. Lett.*, A7, 467 (1992).

233 [7] The inverse dependence on the mass of the neutrino is reasonable because no proton decays at any finite energy are possible for zero mass neutrinos. See R. Ehrlich, Implications for the cosmic ray spectrum of a negative electron neutrino (mass)^2, *Phys. Rev.*, D60, 17302 (1999), for the derivation.

[8] V. A. Kostelecky, in F. Mansouri and J. J. Scanio, eds., *Topics on Quantum Gravity and Beyond* (Singapore: World Scientific, 1993).

[9] Ehrlich, Implications for the cosmic ray spectrum; also see http://xxx.lanl.gov/abs/astro-ph/9812336.

[10] R. Ehrlich, Is there a 4.5 PeV neutron line in the cosmic rays? *Phys. Rev.*, D60, 73005 (1999); also see http://xxx.lanl.gov/abs/astro-ph/9904290.

[11] For references on the experiments on neutral particles from Cygnus X-3, see ibid.

[12] J. Lloyd-Evans et al., Observations of γ rays > 10^{15} eV from Cygnus X-3, *Nature*, 305, 784 (1983).

[13] See J. W. Cronin et al., Search for discrete sources of 100 TeV gamma radiation, *Phys. Rev.*, D45, 4385 (1992), and A. Borione et al., High statistics search for ultrahigh energy γ-ray emission from Cygnus X-3 and Hercules X-1, *Phys. Rev.*, D55, 1714 (1997).

Chapter 10. There Was No Big Bang

[1] Some supernovas emit so much light during their brief outburst that they outshine their entire galaxy of 100 billion stars.

[2] G. Burbidge, F. Hoyle, and J. V. Narlikar, A different approach to cosmology, *Phys. Today*, 52, 38–44 (1999), and A. Albrecht, Reply to "A different approach to cosmology," *Phys. Today*, 52, 44–46 (1999). Also see *A Different Approach to Cosmology*, (Cambridge, UK: Cambridge University Press, 2000), by the same authors.

[3] B. Burbidge, The reality of anomalous redshifts in the spectra of some QSOs and its implications, *Astron. Astrophys.*, 309, 9–22 (1996); G. Burbidge, A. Hewitt, J. V. Narlikar, and P. Das Gupta, Associations between quasistellar objects and galax-

ies, *Ap. J. Suppl.*, 74, 675–730 (1990); G. Burbidge, Redshifts and distances, *Nature*, 282, 451–55 (1979).

[4] R. Webster et al., Detection of statistical gravitational lensing by foreground mass distributions, *Nature*, 336, 358–59 (1988).

[5] C. Canizares, Not too close for comfort, *Nature*, 336, 309–10 (1988).

[6] J. R. Gott III, A time-symmetric, matter, antimatter, tachyon cosmology, *Ap. J.*, 187, 1–3 (1987).

Bibliography

Alvarez, W. *T. rex and the Crater of Doom*. Princeton University Press, Princeton, NJ, 1997.

Cole, L. A. *Element of Risk: The Politics of Radon*. Oxford University Press, Oxford, UK, 1993.

Derry, G. N. *What Science Is and How It Works*. Princeton University Press, Princeton, NJ, 1999.

Duesberg, P. H. *Inventing the AIDS Virus*. Regnery Publishing, Washington, DC, 1996.

Duesberg, P. H. *Stretching the Germ Theory beyond Its Limits*. North Atlantic Books, Berkeley, CA, 1996.

Duesberg, P. H. and J. Yiamouyiannis. *AIDS: The Good News Is HIV Doesn't Cause It*. Health Action Press, Dordrecht, The Netherlands, 1995.

Duesberg, P. H., ed. *AIDS: Virus or Drug Induced?* Kluwer Academic Publishers, Dordrecht, The Netherlands, 1996.

Dyson, F. *The Sun, the Genome, and the Internet: Tools of Scientific Revolutions*. Oxford University Press, Oxford, UK, 1999.

Dyson, F. *Imagined Worlds*. Harvard University Press, Cambridge, MA, 1998.

Dyson, F. *Origins of Life*. Cambridge University Press, Cambridge, UK, 1999.

Ehrlich, R. *The Cosmological Milkshake*. Rutgers University Press, New Brunswick, NJ, 1994.

Ehrlich, R. *What If You Could Unscramble an Egg?* Rutgers University Press, New Brunswick, NJ, 1996.

Ehrlich, R., *What If? Mind-Boggling Science Questions for Kids*. John Wiley and Sons, New York, 1998.

Einstein, A. *Relativity: The Special and the General Theory*. Three Rivers Press, New York, 1961.

Gallo, R. *Virus Hunting: AIDS, Cancer, and the Human Retrovirus: A Story of Scientific Discovery*. Basic Books, New York, 1991.

Geroch, R. *General Relativity from A to B*. University of Chicago Press, Chicago, 1981.

Glashow, S. L. *From Alchemy to Quarks: The Study of Physics as a Liberal Art*. Brooks/Cole Publishing Company, Pacific Grove, CA, 1994.

Gold, T. *The Deep Hot Biosphere*. Springer-Verlag, New York, 1999.

Gold, T. *Power from the Earth*. J. M. Dent and Sons, London, 1987.

Gould, S. J. *The Mismeasure of Man*. W. W. Norton, New York, 1996.

Gribben, J. *Schrödinger's Kittens and the Search for Reality*. Little, Brown, Boston, 1995.

Guth, A. *The Inflationary Universe*. Addison-Wesley Longman, Reading, MA, 1997.

Hawking, S., and Penrose, R. *The Nature of Space and Time*. Princeton University Press, Princeton, NJ, 1996.

Hawking, S. *A Brief History of Time*. Bantam Doubleday Dell, New York, 1998.

Hazen, R. *Why Aren't Black Holes Black? The Unanswered Questions at the Frontiers of Science*. Doubleday, New York, 1997.

Hellman, H. *Great Feuds in Science: Ten of the Liveliest Disputes Ever*. John Wiley and Sons, New York, 1998.

Houghton, J. *Global Warming: The Complete Briefing*. Cambridge University Press, Cambridge, UK, 1994.

Humphrey, N. *Leaps of Faith: Science, Miracles, and the Search for Supernatural Consolation*. Basic Books, New York, 1996.

Kondo, S. *Health Effects of Low-Level Radiation*. Kinki University Press, Osaka, Japan; and Medical Physics Publishing, Madison, WI, 1993.

Kragh, H. *Cosmology and Controversy: The Historical Development of Two Theories of the Universe*. Princeton University Press, Princeton, NJ, 1996.

Lemonick, M. D. *Light at the Edge of the Universe: Dispatches from the Front Lines of Cosmology*. Princeton University Press, Princeton, NJ, 1995.

Lewis, J. S. *Rain of Iron and Ice: The Very Real Threat of Comet and Asteroid Bombardment*. Addison-Wesley, Reading, MA, 1996.

Lott Jr., J. *More Guns, Less Crime*. University of Chicago Press, Chicago, 1998.

Mermin, N. D. *Space and Time in Special Relativity*. Waveland Press, Prospect Heights, IL, 1989.

Mendelsohn, R., and Neumann, J. E., eds. *The Impact of Climate Change on the United States Economy.* Cambridge University Press, Cambridge, UK, 1999.

Nahin, P. J. *Time Machines: Time Travel in Physics, Metaphysics, and Science Fiction.* American Institute of Physics Press, New York, 1993.

Novikov, I. D. *The River of Time.* Cambridge University Press, Cambridge, UK, 1998.

Parks, R. L. *Voodo Science: The Road from Foolishness to Fraud.* Oxford University Press, New York, 2000.

Parsons, M. L., and Singer, S. F. *Global Warming: The Truth behind the Myth.* Perseus Publishing, Cambridge, MA, 1995.

Philander, S. G. *Is the Temperature Rising? The Uncertain Science of Global Warming.* Princeton University Press, Princeton, NJ, 1998.

Powell, J. L. *Night Comes to the Cretaceous: Dinosaur Extinction and the Transformation of Modern Geology.* W. H. Freeman, New York, 1996.

Shermer, M., *Why People Believe Weird Things: Pseudoscience, Superstition, and Other Confusions of Our Time,* W. H. Freeman, New York, 1998.

Singer, S. F. *Hot Talk, Cold Science: Global Warming's Unfinished Debate.* The Independent Institute, Oakland, CA, 1998.

Spence, P., ed. *The Universe Revealed.* Cambridge University Press, Cambridge, UK, 1998.

Thorne, K. S., *Black Holes and Time Warps.* W. W. Norton, New York, 1994.

Taylor, E. F., and Wheeler, J. A. *Spacetime Physics.* 2nd ed. W. H. Freeman, New York, 1992.

Trefil, J. S. *From Atoms to Quarks: An Introduction to the Strange World of Particle Physics.* Anchor Books, New York, 1994.

Trefil, J. S. *Edge of the Unknown: 101 Things You Don't Know about Science and No One Else Does Either.* Houghton Mifflin, Boston, 1996.

Trefil, J. S., and Hazen, R. M. *Sciences: An Integrated Approach.* John Wiley and Sons, New York, 2000.

Index